Springer Aerospace Technology

The series explores the technology and the science related to the aircraft and spacecraft including concept, design, assembly, control and maintenance. The topics cover aircraft, missiles, space vehicles, aircraft engines and propulsion units. The volumes of the series present the fundamentals, the applications and the advances in all the fields related to aerospace engineering, including:

- structural analysis,
- aerodynamics,
- aeroelasticity,
- aeroacoustics,
- flight mechanics and dynamics
- orbital maneuvers,
- avionics,
- systems design,
- materials technology,
- launch technology,
- payload and satellite technology,
- space industry, medicine and biology.

The series' scope includes monographs, professional books, advanced textbooks, as well as selected contributions from specialized conferences and workshops.

The volumes of the series are single-blind peer-reviewed.

To submit a proposal or request further information, please contact: Mr. Pierpaolo Riva at pierpaolo.riva@springer.com (Europe and Americas) Mr. Mengchu Huang at mengchu.huang@springer.com (China)

The series is indexed in Scopus and Compendex

More information about this series at https://link.springer.com/bookseries/8613

Huafeng Ding
Editor

Aerospace Mechatronics and Control Technology

Selected Contributions from 2021 7th Asia Conference on Mechanical Engineering and Aerospace Engineering

Springer

Editor
Huafeng Ding
School of Mechanical Engineering
and Electronic Information
China University of Geosciences
Wuhan, Hubei, China

ISSN 1869-1730 ISSN 1869-1749 (electronic)
Springer Aerospace Technology
ISBN 978-981-16-6639-1 ISBN 978-981-16-6640-7 (eBook)
https://doi.org/10.1007/978-981-16-6640-7

This Springer imprint is published by the registered company Springer Nature Singapore Pte Ltd.
The registered company address is: 152 Beach Road, #21-01/04 Gateway East, Singapore 189721, Singapore

Preface

This proceeding records the papers presented at 2021 7th Asia Conference on Mechanical Engineering and Aerospace Engineering (MEAE 2021), which was held on June 11–13, 2021 in Wuhan, China. It's organized by China University of Geosciences (Wuhan) and hosted by School of Mechanical Engineering and Electronic Information of CUG (Wuhan).

The success of the conference is inseparable from the unremitting efforts of a dedicated conference team and the strong support of China University of Geosciences. We invite all successfully registered authors and listeners to join us, interact with keynote speakers and invited experts, and most importantly, interact with all fellow participants. Without the hard work of many volunteers and the organizing committee, there would be no such high-quality meeting. In particular, we would like to express our sincere thanks to the outstanding session chair team and TPC members.

The success of MEAE 2021 depends on the contributions of many individuals and organizations. With that in mind, we thank all authors who submitted work to the conference. The quality of submissions this year remains high and we are satisfied with the quality of the results procedure. The organizing committee also thanked the members of the Technical Planning Committee and the chairman of the meeting for their strong support. The organizing committee is responsible for reviewers, who voluntarily sacrifice valuable time to evaluate the manuscript and provide authors with useful feedback.

Finally, we hope that participants will like this conference and have the opportunity to establish contacts with colleagues from all over the world. We also hope that all participants can learn from this conference and contribute more to your future exploration and research.

Best wishes,

Wuhan, China

MEAE Conference Committee

Contents

Aircraft Control and Testing

Numerical Study on Aerodynamic Performance of Hypersonic Vehicle with Aerospikes

Yang Xu(✉), Shu Zhou Fang, and Shuai Zhang

School of Aerospace and Engineering, Beijing Institute of Technology, Beijing 100081, China
1773123413@qq.com, fsz@bit.edu.cn

Abstract. The drag and heat flux reduction have become an increasing crucial characteristic for the hypersonic vehicles. Numerical simulations of the hypersonic flow around the blunt body with aerospike are investigated for the aerospike's changing *L/D* ratios with freedom Mach number of 5.75. The effect of the characteristics of the flow field and the aerospike's *L/D* ratios on the drag coefficient of model and the surface pressure coefficient of the nose cone are analysed using a two-dimensional axisymmetric Reynolds-averaged Navier–Stokes (RANS) equations coupled with the two equation k-ω shear stress transport (SST) turbulence model. The numerical simulation results show that the installation of the aerospike changes the flow field structure of the blunt body vehicle head, which reduces the drag of the blunt body vehicle significantly in hypersonic flight. The size of the recirculation region depends on the aerospike's *L/D* ratios. Among the different aerospike configurations studied, the drag reduction effect of aerospike when the $L/D = 1$ is the best, and the maximum value of the drag reduction efficiency is 36.4%. Within the scope of this study, as the *L/D* ratio of the aerospike increases, the drag reduction efficiency increases gradually.

Keywords: Hypersonic flow · Aerospike · Drag reduction

1 Introduction

Hypersonic vehicle will encounter serious aerodynamic drag and aerodynamic heating when it re-enters the atmosphere. The severe aerodynamic heating phenomenon will damage the internal electronic equipment of the vehicle, and the huge shock wave drag also hinders further improvement of the vehicle speed. Hence, the study of heat flux and drag reduction of hypersonic vehicle is of great significance to the hypersonic vehicle's aerodynamic performance and safety performance.

A great deal of thermal protection methods and drag reduction techniques were proposed and studied in detail. For example, the forward-facing cavity [1], counterflowing jet [2], the energy deposition [3] and the different combinations of them [4]. Besides the methods above, attaching an aerospike to the head of the vehicle which flights at hypersonic speed appears to be a simplest technique which also has a higher efficiency. The experimental investigations of the aerospike structure began in the 1950s.

Kalimuthu et al. [5] studied the effects of hemispherical model with aerospike on drag reduction, and the parameters of model in detail. The experimental data indicate

H. Ding (ed.), *Aerospace Mechatronics and Control Technology*, Springer Aerospace Technology,
https://doi.org/10.1007/978-981-16-6640-7_1

that aerodisk with suitable length-to-diameter ratios and head structure may be able to decrease aerodynamic drag. Ahmed and Qin [6] studied the effects of aerospike of variable length and aerodisk of variable size on vehicle drag reduction characteristics under hypersonic conditions by numerical simulation. The results from the numerical simulation indicate the optimal size of aerodisk produces the minimum drag, and the optimal size is inversely proportional to the length of the aerospike. Gerdroodbary and Hosseinalipour [7] used numerical methods to investigate the aerodynamic performance of aerospike of different lengths and different shapes of aerodisk for changing angles of attack. The data indicate that blunt body models' surface heat flux is reduced meanwhile blunt body models' surface pressure coefficient is also reduced which have different types of aerospike, and the size of recirculation region is different among different models. Ahmed and Qin [8] studied the aerodynamic performance of spiked blunt bodies under the condition of Ma = 6. The numerical results indicate that the installation of the aerodisk reduces the aerodynamic heat protection performance of the aerospike, but its heat protection performance is still better than that of the blunt body model without aerospike.

Mansour and Khorsandi [9] studied the flow near the blunt body model with aerospike under hypersonic conditions. The numerical results show that the model's drag coefficient with aerospike descend by 40% than the model which has no aerospike, and the surface pressure coefficient of the model obtained from numerical calculation is highly consistent with the relevant experimental data. Sebastian et al. [10] researched hypersonic aerodynamic performance of aerospike for changing angles of attack. The results which from the numerical simulation are similar to the results from the relevant experimental investigation. Deng et al. [11] studied the lifting body's flight aerodynamic characteristics for one model with aerospike, and the other models have no aerospike at Ma = 8. The numerical results show that the model of $L/D = 2$ has the optimal drag reduction efficiency. Huang et al. [12] researched the features of flow field of blunt body model with and without aerospike under hypersonic conditions, and studied the mechanism of aerodisk diameter and aerospike length in drag and heat flux reduction. Numerical data show that the dynamic pressure highly decreases in recirculation region which makes the surface drag and heat flux of the model reduce. Narayan et al. [13] studied the aerospike with different geometric configurations by numerical and experimental methods for finding out the optimal modified geometry. Both of the numerical and experiment results show that the geometry with a stepped aerospike has the optimal performance within the scope of the research for decreasing aerodynamic drag and aerodynamic heating.

2 Geometry Model and Numerical Method

2.1 Geometry Model

Figure 1 gives the schematic diagram of the geometry model. The aerospike is installed at the head of the hypersonic blunt model, with the geometry model's center as the coordinate origin. The diameter of blunt model (D) is 50 mm meanwhile the diameter of aerospike (d) is 4 mm. The aerospike's length (L) is variable for studying the effects of L/D ratios on drag reduction efficiency.

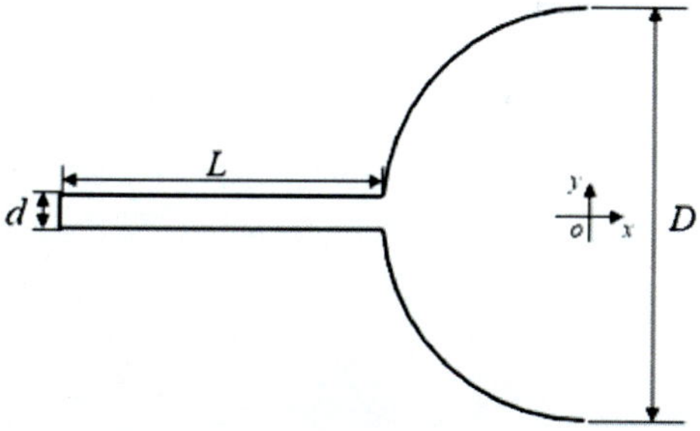

Fig. 1. Schematic diagram of the geometry

2.2 Numerical Method

The data of numerical simulation have been gained using ANSYS FLUENT CFD version 19.2 under the steady state conditions. In this work, two-dimensional axisymmetric Reynolds-Averaged Navier–Stokes (RANS) equations are chosen to numerical calculations meanwhile the SST k-ω turbulence model is selected to simulate the viscosity characteristics for simulating the features of flow field near the aerospike model. At the same time, the double precision solver is employed to solve the equations.

The air is considered as the perfect gas, besides the Sutherland's viscosity law is used. For accelerating the convergence speed of the calculation process, the first order upwind scheme and the advection upstream splitting method (AUSM) has been selected. On the other hand, the Courant number is set as 0.5 [14]. When all of the residual values fall for not less than six orders of magnitude meanwhile difference of mass flow rate between the calculated inflow and outflow decrease under 0.001 kg/s, the data are assumed to be convergent.

2.3 Computational Grids and Boundary Conditions

The structural grids have been obtained using commercial software ANSYS ICEM CFD version 19.2. The schematic diagram of the grid division is showed in Fig. 2. In order to meet the requirements of the boundary layer, all the walls' height of the first row of cells is set to 1×10^{-5} m, and the grid density of all the grids is gathered towards the wall. The Fig. 3 and Table 1 give the information of boundary conditions for the CFD simulation in detail.

3 Numerical Method Validation and Grid Independency Analysis

3.1 Numerical Method Validation

In this section, the data from numerical simulation was compared with experimental results that measured by Kalimuthu [16]. Figure 4 depicts the pressure coefficient distribution for the hemispherical body with cylinder afterbody. Figure 5 depicts the Mach number contour obtained by simulation and experiment schlieren picture of Kalimuthu [16]. The numerical data are very similar to the experimental results, then biggest difference appears in the vicinity of the stagnation point.

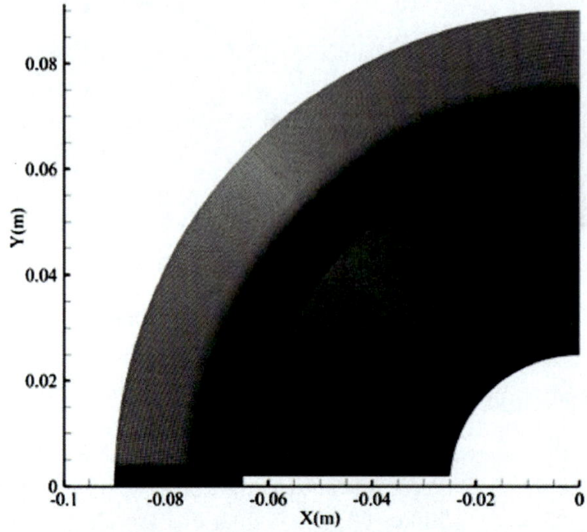

Fig. 2. Computing grids division

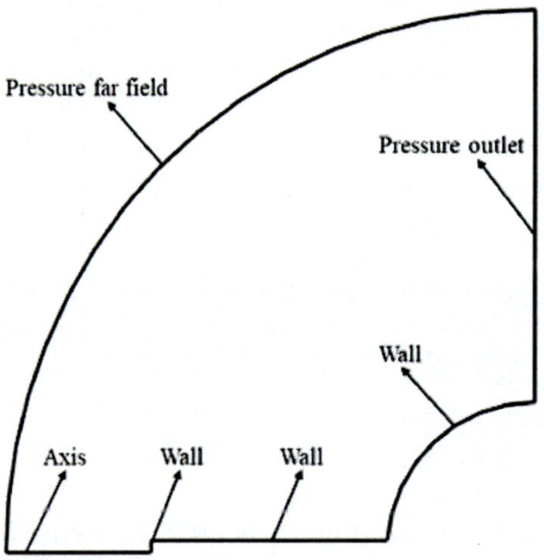

Fig. 3. Boundary conditions and domain

3.2 Grid Independency Analysis

For the stability and accuracy of simulation computation, grid quality is very critical. Hence, a grid-independent study of the model of $L/D = 0.6$ has been performed. In this section, three different grid scales with 150,638, 230,298 and 336,778 grids are generated. Figures 6 and 7 depict the surface stanton number distribution and surface pressure coefficient distribution for the grid divisions above, respectively. Both of the values and the changing trends of stanton number and pressure coefficient of the three grid scales are roughly similar. Considering the saving of calculation cost, the moderate grid is selected for following study.

Table 1. Boundary conditions of the numerical simulation [15]

Part	Condition	Parameter	Value
Inlet	Pressure far field	Mach Number	5.75
		Pressure (Pa)	425
		Temperature (K)	140
Outlet	Pressure outlet	Pressure (Pa)	0.00001
		Temperature (K)	295
Body Head	Wall	Temperature (K)	300
Spike Head	Wall	Temperature (K)	300
Spike Up	Wall	Temperature (K)	300
Axis	Axis		

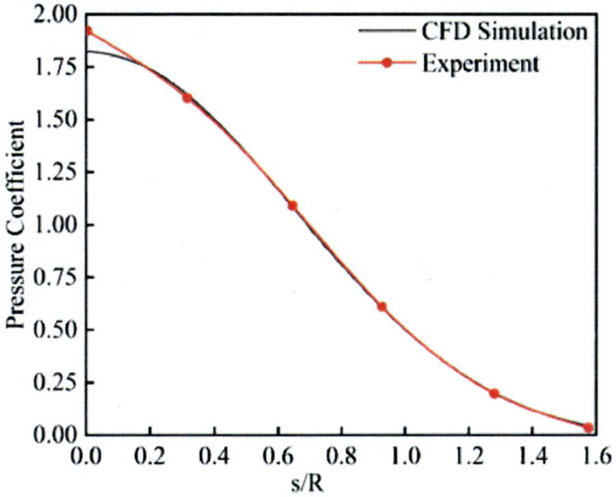

Fig. 4. Pressure coefficient distribution

4 Computational Results and Discussion

In the current study, six different examples are set up to study the influence of changing *L/D* ratios of aerospike on drag reduction efficiency of blunt body. Table 2 shows the setting of six different examples, where example 1(i.e. $L = D = 0$ mm) means that there is no aerospike structure.

4.1 Characteristics of Flow Field

The simulation of hypersonic flow is performed for the blunt body model with and without aerospike. Figure 8 depicts the Mach number contours of the model with aerospike which has changing *L/D* ratios. From figure we can see clearly that the bow shock wave

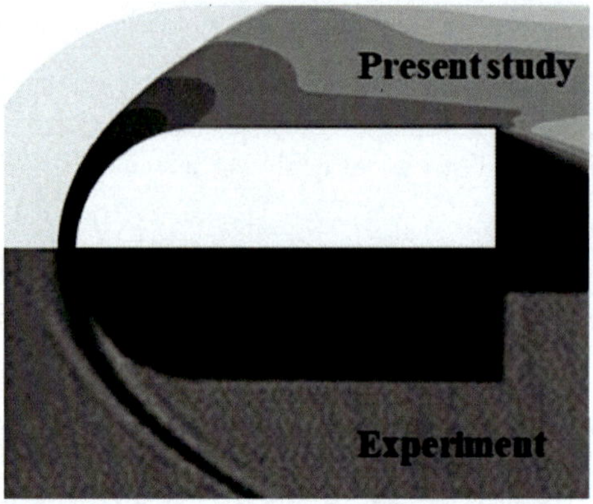

Fig. 5. Mach contour and schlieren picture

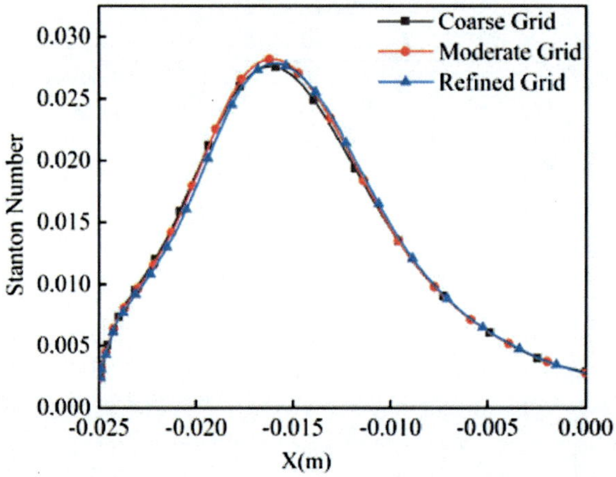

Fig. 6. Stanton number distribution

is converted into an oblique shock wave at the head of blunt body by the aerospike, which produces lower pressure and temperature than the bow shock wave. The structure of flow field in vicinity of head of the model has been changed by the installation of the aerospike, which makes the flow's separation and reattachment. At a certain distance from the oblique shock wave's downstream, the boundary layer on the surface of the aerospike separates, forming a shear layer and propagating downstream, and then the shear layer is reattached in vicinity of the shoulder of the model to form a reattachment shock wave. The reattachment shock wave produced by the head of the model interacts with the aerospike, which makes the oblique shock wave generate.

Figure 9 depicts the temperature contours and streamline of the model with aerospike with changing *L/D* ratios. From figure we can get that the aerospike produces a recirculation separation flow region of low temperature and low pressure, which protects the model from the incoming flow. Near downstream of the head of the aerospike up to the model shoulder's flow reattachment point, the recirculation region begins to form. The

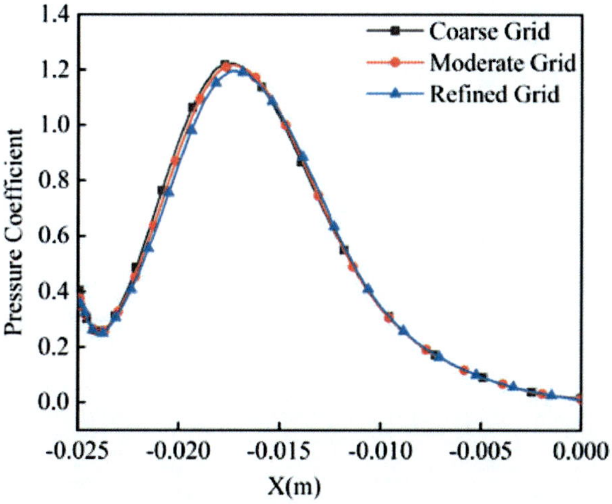

Fig. 7. Pressure coefficient distribution

Table 2. Setting of different examples

Examples	1	2	3	4	5	6
L (mm)	0	10	20	30	40	50
D (mm)	0	50	50	50	50	50
L/D		0.2	0.4	0.6	0.8	1.0

temperature near the reattachment point reaches a higher value. The aerospike's length has a great influence on recirculation region's shape, and recirculation region's size is proportional to the aerospike's *L/D* ratios. Therefore, if we want to obtain the maximum value of the drag reduction efficiency, the aerospike's *L/D* ratios must be appropriately selected to gain a larger conical recirculation region upstream of the nose cone.

4.2 Influences of *L/D* Ratios on Drag

The drag coefficient distribution of the model with varying *L/D* ratios of aerospike is depicted in Fig. 10, including the drag coefficient distribution of the model without aerospike. Compared with the blunt body which has no aerospike, the drag coefficient is reduced by introducing the aerospike, and with the increase of the aerospike's *L/D* ratios, the drag coefficient is gradually smaller.

Table 3 shows the values of drag coefficient comparison for examples with changing *L/D* ratios. Δ is drag reduction coefficient. The formula of Δ is as follows.

$$\Delta = \frac{C_d - C_{dref}}{C_{dref}} \times 100\% \tag{1}$$

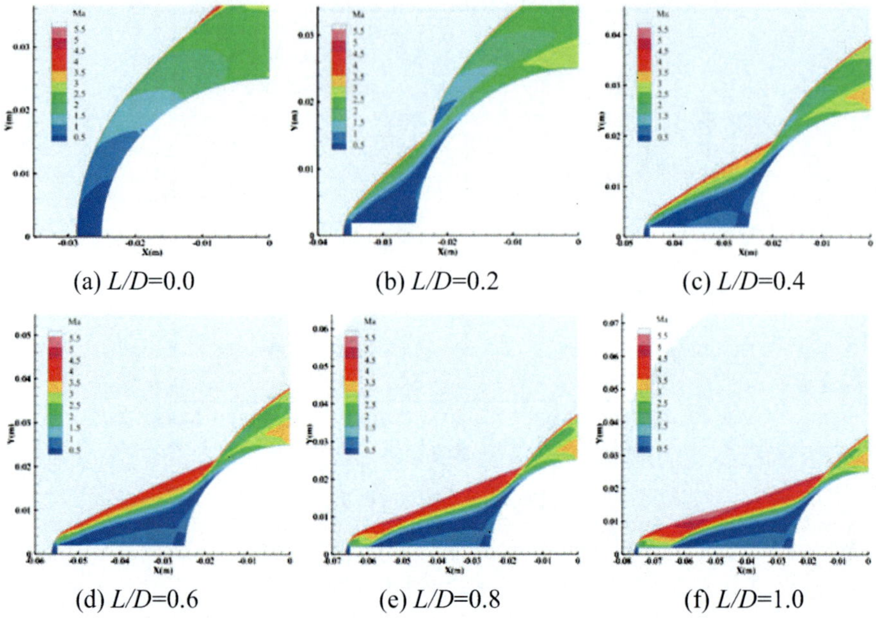

(a) L/D=0.0 (b) L/D=0.2 (c) L/D=0.4

(d) L/D=0.6 (e) L/D=0.8 (f) L/D=1.0

Fig. 8. Mach number contours for the changing L/D ratios

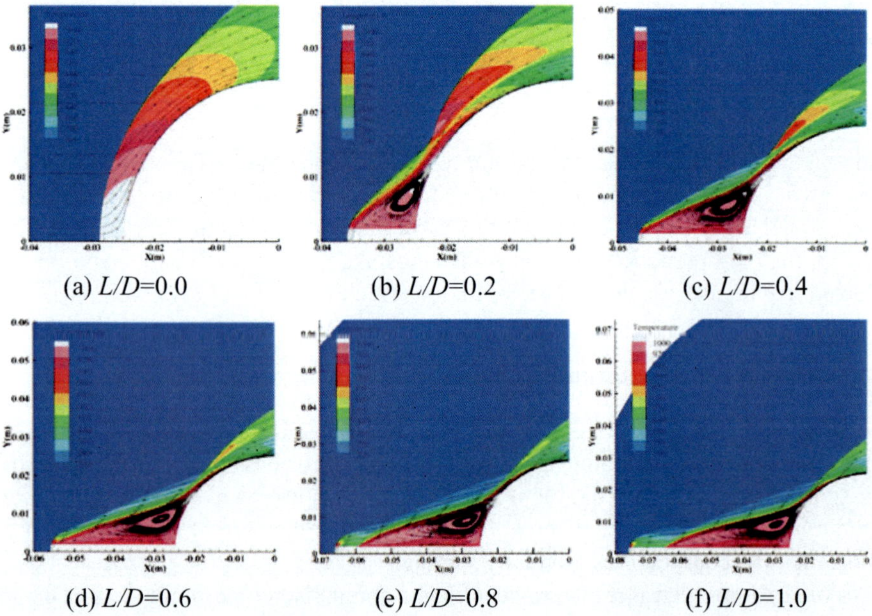

(a) L/D=0.0 (b) L/D=0.2 (c) L/D=0.4

(d) L/D=0.6 (e) L/D=0.8 (f) L/D=1.0

Fig. 9. Temperature contours and streamline comparison for different L/D ratios

where, C_d is the drag coefficients for the different examples which have the aerospikes, and C_{dref} is the model's drag coefficient which has no aerospike.

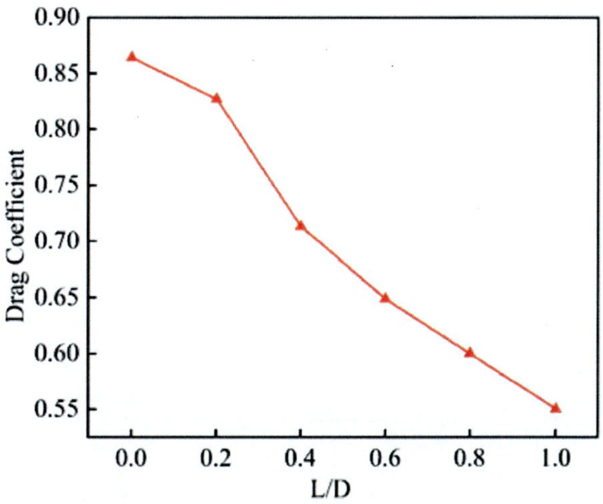

Fig. 10. Drag coefficient distribution of different *L/D* ratios

Table 3. Drag coefficient comparison of examples with different *L/D* ratios

L/D	C_d	Δ (%)	*L/D*	C_d	Δ (%)
0.0	0.86438	0	0.6	0.64903	−24.9
0.2	0.82711	−4.3	0.8	0.60006	−30.6
0.4	0.71336	−17.5	1.0	0.55015	−36.4

The model's drag coefficient is 0.86438 which has no aerospike. The models' drag coefficients which have aerospikes are lower than that the model's drag coefficient which has no aerospike. The *L/D* ratios of aerospike are inversely proportional to the drag coefficient. When $L/D = 1.0$ of the aerospike, the value of drag coefficient reaches the minimum, and the maximum drag reduction efficiency is 36.4%.

4.3 Effects of *L/D* Ratio on Pressure

Figure 11 depicts the pressure coefficient distribution of the model with aerospike's changing *L/D* ratios, including the pressure coefficient distribution of the model without aerospike. From the figure we can see the pressure coefficient's maximum value of the model without aerospike appears at the stagnation point of the geometry head, then along the surface of the blunt body decreases gradually, indicating that the flow continues to accelerate. The huge stagnation pressure on the head caused a lot of drag. The installation of the aerospike changes the flow field's features and reduces the pressure coefficient's maximum value.

For the model with aerospike, the pressure coefficient increases first of all, then decreases, and reaches the peak in vicinity of the shoulder of the model. The peak position of the pressure coefficient, that is, the region near the reattachment point, with

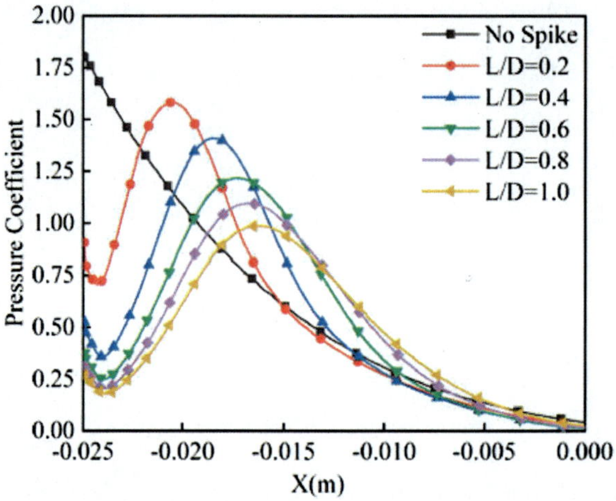

Fig. 11. Pressure coefficient distribution for different *L/D* ratios

the aerospike's *L/D* ratios continuous increase, the position of the maximum value of pressure coefficient continues to move backward, that is, the reattachment point's position gradually moves backward, leading to the recirculation region's size continuous increase, so drag reduction efficiency increases gradually, which is consistent with the above analysis results.

5 Conclusions

Numerical simulations of the hypersonic flow in vicinity of the blunt body model with aerospike are investigated for changing *L/D* ratios of the aerospike. The numerical results show that:

(1) The installation of the aerospike changes the flow field structure of the hypersonic vehicle nose cone and forms the recirculation region of low speed, low temperature and low pressure, thus significantly reducing the drag of the vehicle in the process of the hypersonic flight.
(2) The recirculation region's size depends on the aerospike's *L/D* ratios, which is the main reason for drag reduction. In the aerospike configuration with changing *L/D* ratios, the drag reduction efficiency of the aerospike with $L/D = 1.0$ is the optimal, and the maximum value of the drag reduction efficiency is 36.4%.
(3) Within the scope of this study, with the increase of the aerospike's *L/D* ratios, the drag reduction efficiency increases gradually. In order to make full use of the drag reduction performance of the aerospike, the shear layer's reattachment point on the surface of the model should be moved backward by selecting the appropriate aerospike's *L/D* ratios to form a larger recirculation region.

References

1. Huang W, Zhao Z-t, Yan L et al (2017) Parametric study on the drag and heat flux reduction mechanism of forward-facing cavity on a blunt body in supersonic flows. J Aerosp Sci Technol 71 619–626
2. Shen B, Liu W, Yin L (2018) Drag and heat reduction efficiency research on opposing jet in supersonic flows. J Aerosp Sci Technol 77:696–703
3. Ashwin Ganesh M, John B (2018) Concentrated energy addition for active drag reduction in hypersonic flow regime. J. Acta Astronautica 142:221–231
4. Sun X, Huang W, Ou M et al (2019) A survey on numerical simulations of drag and heat reduction mechanism in supersonic/hypersonic flows. J Chinese J Aeronaut 32(4):771–784
5. Kalimuthu R, Mehta RC, Rathakrishnan E (2008) Experimental investigation on spiked body in hypersonic flow. J. Aeronaut J 112(1136):593–598
6. Ahmed MYM, Qin N (2010) Drag reduction using aerodisks for hypersonic hemispherical bodies. J Spacecr Rockets 47(1):62–80
7. Gerdroodbary MB, Hosseinalipour SM (2010) Numerical simulation of hypersonic flow over highly blunted cones with spike. J Acta Astronaut 67(1):180–193
8. Ahmed MYM, Qin N (2011) Numerical investigation of aeroheating characteristics of spiked blunt bodies at Mach six flight conditions. J Aeronaut J 115(1168):377–386
9. Mansour K, Khorsandi M (2014) The drag reduction in spherical spiked blunt body. J Acta Astronaut 99:92–98
10. Sebastian JJ, Suryan A, Kim HD (2016) Numerical analysis of hypersonic flow past blunt bodies with aerospikes. J Spacecr Rockets 53(4):669–677
11. Deng F, Jiao Z, Liang B et al (2017) Spike effects on drag reduction for hypersonic lifting body. J Spacecr Rockets 54(6):1185–1195
12. Huang W, Li L-q, Yan L et al. Drag and heat flux reduction mechanism of blunted cone with aerodisks, J. Acta Astronautica. 138, (2017) 168–175
13. Narayan A, Narayanan S, Kumar R et al (2019) Control of aerodynamic drag and heating of nose cones through taper spikes. J Spacecr Rockets 56(4):1165–1176
14. Huang W, Li L-q, Chen X-q et al (2017) Parametric effect on the flow and mixing properties of transverse gaseous injection flow fields with streamwise slot: A numerical study. Int J Hydrogen Energy 42(2): 1252–1263
15. Ou M, Yan L, Huang W et al (2019) Design exploration of combinational spike and opposing jet concept in hypersonic flows based on CFD calculation and surrogate model. J Acta Astronaut 155:287–301
16. Kalimuthu R, Rathakrishnan E. Aerospike for drag reduction in hypersonic flow. In: C. 44th AIAA Joint Propulsion Conference & Exhibit, Hartford, CT, 21–23 July 2008

Terminal Multi-Constrained Finite Time Sliding Mode Guidance Law Based on Truncation Function for Hypersonic Vehicle

Shenghui Cui(✉), Cheng Hu, Xinbin Bai, and Shifeng Zhang

National University of Defense Technology, College of Aerospace Science and Engineering, Changsha 710073, China
745032917@qq.com

Abstract. In order to solve the terminal guidance problem of the hypersonic vehicle, a multi-constrained guidance law with position, impact angle and impact velocity was studied. Firstly, a finite time sliding mode control method based on truncation function for second order nonlinear system was proposed, and then a reasonable truncation function was designed according to the dynamic equation of the terminal guidance phase of the vehicle to form the sliding mode control function of the longitudinal and transverse planes. Further, the auxiliary control was obtained and the guidance command was generated. Under the action of the control quantities, the system was kept on the sliding mode surface all the time, that is, the global sliding mode can be realized, and the constraints of position and impact angle can be realized. The guidance parameter was solved iteratively according to the redundant degrees of freedom in the transverse plane to ensure the constraint of impact velocity. The simulation verified that the guidance method can meet the requirements of terminal multiple constraints of hypersonic vehicle.

Keywords: Multi-constrained guidance law · Limited time sliding mode control · The truncation function · Terminal velocity constraint · Impact angle constraint

1 Introduction

Guidance system is the brain of missile, which directly determines the success of flight mission. As an important branch of guidance technology, precision guidance technology develops rapidly in recent decades, which brings a revolutionary change to modern military.

The design of terminal guidance method in the reentry attack phase directly determines the success or failure of the attack mission, and it is one of the important parts in the reentry phase of the vehicle. Aiming at the requirement of the velocity angle at the end of attack, Lu [1] decomposed the three-dimensional space of flight trajectory into two orthogonal planes, longitudinal and transverse, based on the proportional method. In the longitudinal and transverse planes, the parameters of guidance low were updated in real

H. Ding (ed.), *Aerospace Mechatronics and Control Technology*, Springer Aerospace Technology,
https://doi.org/10.1007/978-981-16-6640-7_2

time to ensure that the impact angle meets the requirements of longitudinal and transverse. With the development of guidance low, adaptive guidance low gradually comes into people's vision. Domestic Li Huifeng et al. [2] established a closed-loop relationship between the guidance parameters and the flight state, so that guidance parameters can be updated adaptively to meet the requirements of flight state adjustment in longitudinal and transverse planes. Finally, the accuracy of impact point and angle of impact were satisfied. Diao Zhaoshi et al. [3–5] proposed the guidance low design with terminal angle of attack constraint, which well met the requirements of terminal seeker field angle constraint. The above literatures can effectively solve the problems of impact angle and the accuracy of the impact point at the end of the attack. However, the problem that some terminal guidance phase need to control the velocity has not been solved effectively.

Based on this question, Ohlmeyer et al. [6] proposed a terminal guidance method which takes the weighted residual flight time as the performance index function to control terminal velocity. He Jianxue et al. [7] proposed a method to control the falling velocity of the reentry maneuvering warhead. The idea of this method is as follows: Firstly, an ideal velocity curve was designed, and then the actual flight velocity was as close as possible to the ideal velocity curve by introducing additional angle of attack, thus realizing terminal velocity control. Xie Daocheng et al. [8] further considered the attack angle constraint and obtained the guidance law with impact angle and velocity constraints. However, the control accuracy of this guidance method is not high, especially in the presence of state error and external disturbance, there is a large deviation between the terminal impact angle and the impact velocity and the expected values. Yu et al. [9] based on the trajectory shaping method, a terminal guidance law with terminal velocity control was proposed. The guidance law can control the vehicle to attack the fixed target on the ground at the expected terminal velocity from the near vertical direction, and ensure that the overload was close to zero at the impact time. The guidance law is robust to wind disturbance and atmospheric density perturbation, and has high terminal accuracy. However, the guidance strategy can only ensure that the vehicle can attack the target at a nearly vertical angle, and cannot control the terminal track yaw angle.

In order to solve the guidance law with impact velocity and impact angle constraints, in this paper, a finite-time guidance law based on truncation function was proposed. It is mainly aimed at second-order nonlinear systems, the robustness and convergence of the closed-loop control system were verified by Lyapunov function. Then, by designing the truncation function, a finite-time sliding mode guidance law was proposed for hypersonic vehicle, which can control the impact velocity under the constraint of flight path angle and flight path azimuth angle.

2 Description of Guidance Problems

2.1 Dynamics Equation of the Reentry Vehicle

The dynamic equation of the vehicle in the trajectory coordinate system is as follows [10]:

$$\begin{cases} \dot{x} = V \cos\gamma \cos\chi \\ \dot{y} = V \sin\gamma \\ \dot{z} = -V \cos\gamma \sin\chi \\ \dot{V} = -\dfrac{D}{m} - g \sin\gamma \\ \dot{\gamma} = \dfrac{L\cos\nu}{mV} - \dfrac{g\cos\gamma}{V} \\ \dot{\chi} = -\dfrac{L\sin\nu}{mV\cos\gamma} \end{cases} \tag{1}$$

where, $[x, y, z]^{\mathrm{T}}$ represents the space position of the vehicle; V is the velocity vector; γ and χ are flight path angle and flight path azimuth angle respectively; ν is angle of bank; L and D are respectively lift and drag; m is the mass of the vehicle. In this paper, the mass of vehicle is taken as 1000 kg, namely m=1000 kg.

2.2 Aerodynamic Model of Hypersonic Vehicle

The lift and drag expressions are as follows [11]:

$$\begin{cases} D = \dfrac{1}{2}C_D\rho V^2 S_M \\ L = \dfrac{1}{2}C_L\rho V^2 S_M \end{cases} \tag{2}$$

where, C_D and C_L are drag and lift coefficients respectively; ρ is atmospheric density; V is velocity;S_M is the pneumatic reference area, this article takes S_M=2.5 m^2.

2.3 Description of Terminal Multi-Constrained Problem

Considering the attack mission with fixed target (T), (x, y, z) is the position of the missile (M), (x_f, y_f, z_f) is the position of the target point, γ_f and χ_f are the flight path angle and flight path azimuth angle of the terminal respectively. Then the guidance mission with impact angle can be expressed as:

$$\begin{aligned} &\lim_{t\to t_f}(x, y, z) = (x_f, y_f, z_f) \\ &\lim_{t\to t_f}(\gamma, \chi) = (\gamma_f, \chi_f) \end{aligned} \tag{3}$$

where, t_f is the terminal attack time, and $t_{go} = t_f - t$ represents the remaining flight time of the missile. Considering the guidance law designed with the altitude as the independent variable, the guidance mission with the constraint of impact angle can be expressed as:

$$\begin{aligned} &\lim_{y \to y_f} (x, z) = (x_f, z_f) \\ &\lim_{y \to y_f} (\gamma, \chi) = (\gamma_f, \chi_f) \end{aligned} \tag{4}$$

The geometrical relationship of the three-dimensional missile-target is as follows (Fig. 1):

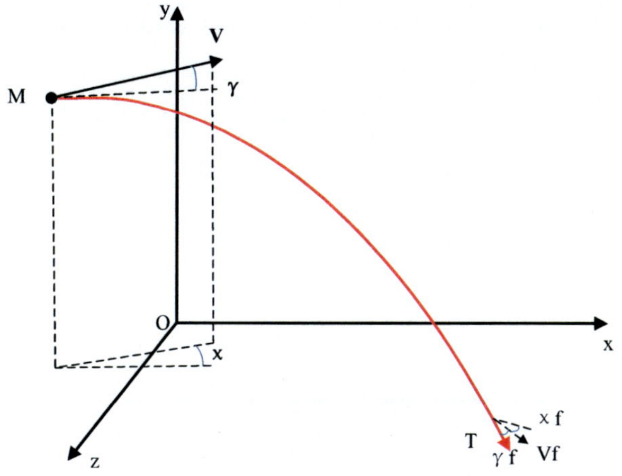

Fig. 1. Geometric relation diagram of missile-target

Taking the longitudinal plane as an example, $z = \chi = 0$. Then the guidance task can be described as:

$$\begin{aligned} &\lim_{y \to y_f} x = x_f \\ &\lim_{y \to y_f} \gamma = \gamma_f \end{aligned} \tag{5}$$

Define a new independent variable with respect to altitude $Y = y_0 - y$, and the derivative of Y is denoted as x', namely:

$$x' = \frac{dx}{dY} = \frac{dx}{dt}\frac{dt}{dY} = -\frac{dx}{dt}\frac{dt}{dy} \tag{6}$$

It can be obtained from (1):

$$x' = -\cot\gamma \tag{7}$$

The significance of the constraint of impact velocity lies in the effective enhancement of missile damage ability. In the process of flight, if the velocity is too high, it may violate the constraints of dynamic pressure, heat flow, overload and so on. If the velocity is too low, it is easy to be intercepted by the other side's anti-missile [12]. Then the constrained guidance task on impact velocity can be described as:

$$\lim_{y \to y_f} V = V_f \tag{8}$$

3 Finite Time Sliding Mode Guidance Law with Impact Angle and Impact Velocity Constraints

3.1 State Variables Design

The state variables about the longitudinal and transverse planes are constructed with reference to the terminal constraint:

$$\begin{aligned} \lambda_1 &= x - x_f + \cot\gamma_f \cos\chi_f (Y - Y_f) \\ \xi_1 &= z - z_f - \cot\gamma_f \sin\chi_f (Y - Y_f) \end{aligned} \tag{9}$$

The derivative is obtained by the above formula with respect to Y:

$$\begin{aligned} \lambda_2 = \lambda_1^{'} &= -\cot\gamma \cos\chi + \cot\gamma_f \cos\chi_f \\ \xi_2 = \xi_1^{'} &= \cot\gamma \sin\chi - \cot\gamma_f \sin\chi_f \end{aligned} \tag{10}$$

It can be seen from (14) and (15) that when the four design variables λ_1, λ_2, ξ_1, ξ_2 converge to zero at the time of impact, the constraints of terminal position and impact angle can be satisfied.

3.2 Guidance Law Design

The sliding mode function of the longitudinal plane is designed as follows:

$$S_1 = \lambda_2 + \frac{n_1\lambda_1}{Y_f - Y} + Q_1 \tag{11}$$

where, $f(t) = Y_f - Y$ is the truncation function and meets the requirement:$Q_1(0) = -\lambda_2(0) - \frac{n_1\lambda_1}{Y_f}$, the derivative can be obtained with respect to Y: $Q_1^{'}(0) = \frac{m_1}{Y_f - Y}(\lambda_2 + \frac{n_1\lambda_1}{Y_f - Y})$. The longitudinal sliding mode function satisfies when $0 \le t \le t_f$, $S_1 \equiv 0$.

The sliding mode function of the transverse plane is:

$$S_2 = \xi_2 - \frac{n_2(2Y + A)}{Y^2 + AY + B}\xi_1 + Q_2 \tag{12}$$

where, $f(t) = Y^2 + AY + B$ is the selection of the truncation function and the parameters A and B are constants. Since $f(t_f) = 0$ is to be satisfied, so $B = -Y_f^2 - AY_f$ can be obtained. Known $Q_2(0) = -\xi_2(0) + \frac{n_2 A}{B}\xi_1(0)$, The derivative is obtained with respect to Y:$Q_2^{'} = -\frac{m_2(2Y+A)}{Y^2+AY+B}(\xi_2 - \frac{n_2(2Y+A)}{Y^2+AY+B}\xi_1)$. The transverse sliding mode function also satisfies when $0 \le t \le t_f$, $S_2 \equiv 0$.

The sliding mode function of the transverse plane has two design variables, so that the impact velocity can be adjusted on the premise of ensuring the constrained flight path azimuth angle. Derivatives of the sliding mode functions in the longitudinal and transverse planes with respect to Y can be obtained as follows:

$$\begin{aligned} S_1^{'} &= \lambda_2^{'} + \frac{(n_1+m_1)\lambda_2}{Y_f - Y} + \frac{(m_1+1)n_1\lambda_1}{(Y_f-Y)^2} \\ S_2^{'} &= \xi_2^{'} - \frac{2n_2(Y^2+AY+B) - n_2(m_2+1)(2Y+A)^2}{(Y^2+AY+B)^2}\xi_1 \\ &- \frac{(n_2+m_2)(2Y+A)}{Y^2+AY+B}\xi_2 \end{aligned} \tag{13}$$

Let $u_1 = \lambda_2^{'}$, $u_2 = \xi_2^{'}$ as the auxiliary control quantities of the system, the following control commands can be obtained:

$$\begin{aligned} u_1 &= -\frac{(n_1+m_1)\lambda_2}{Y_f - Y} - \frac{(m_1+1)n_1\lambda_1}{(Y_f-Y)^2} - k_1 Sat(S_1) \\ u_2 &= \frac{2n_2(Y^2+AY+B) - n_2(m_2+1)(2Y+A)^2}{(Y^2+AY+B)^2}\xi_1 \\ &+ \frac{(n_2+m_2)(2Y+A)}{Y^2+AY+B}\xi_2 - k_2 Sat(S_2) \end{aligned} \tag{14}$$

where, n_1, n_2, m_1, m_2 are constants; k_1, k_2 are the switching gain constants of the longitudinal and transverse planes.$Sat(s)$ is the saturation function,which is considered to replace the switching function in order to avoid chattering [13]. The saturation function is designed as follows:

$$Sat(s) = \begin{cases} \varsigma^{-1}s & |s| \le \varsigma \\ \mathrm{sgn}(s) & |s| > \varsigma \end{cases} \tag{15}$$

where, ς is the thickness of the boundary layer, which directly affects the effect of chattering suppression.

Further derivative of Eq. (10) with respect to Y is obtained as follows:

$$\begin{aligned} \lambda_2^{'} &= \frac{\cos\chi}{\sin^2\gamma}\gamma^{'} + \chi^{'}\cot\gamma\sin\chi \\ \xi_2^{'} &= -\frac{\sin\chi}{\sin^2\gamma}\gamma^{'} + \chi^{'}\cot\gamma\cos\chi \end{aligned} \tag{16}$$

Substitute $u_2 = \xi_2^{'}$,$u_1 = \lambda_2^{'}$ into Eq. (16), and we can get:

$$\begin{aligned} \gamma^{'} &= (u_1\cos\chi - u_2\sin\chi)\sin^2\gamma \\ \chi^{'} &= (u_1\sin\chi + u_2\cos\chi)\tan\gamma \end{aligned} \tag{17}$$

For hypersonic vehicle, the guidance commands formed by the designed guidance law are generally reflected in the control of attack angle α and bank angle ν (BTT). From the dynamics equation of the vehicle in Eq. (1), the expressions of lift force and bank angle can be obtained:

$$L = mV^2 \sin\gamma \sqrt{\left(\frac{g\cos\gamma}{V^2\sin\gamma} - \gamma'\right)^2 + (\chi' cos\gamma)^2} \tag{18}$$

The expression of lift coefficient can be further obtained from the expression of lift force:

$$C_L = \frac{2m\sin\gamma\sqrt{\left(\frac{g\cos\gamma}{V^2\sin\gamma} - \gamma'\right)^2 + (\chi' cos\gamma)^2}}{\rho S_M} \tag{19}$$

where, ρ is atmospheric density.

$$\nu = atan\frac{\cos\gamma\chi'}{\frac{g\cos\gamma}{V^2\sin\gamma} - \gamma'} \tag{20}$$

Substitute (17) into (19) to obtain guidance commands about lift coefficient. Then use linear interpolation and MATLAB optimization algorithm (such as fzeros) to find the real-time angle of attack command.

It can be found that the value of parameter A does not affect the final convergence result, so the parameter A is considered as an extra degree of freedom to constrain the terminal velocity. The selection of parameter A is derived as follows:

Under the action of the designed guidance law, the sliding mode functions of the longitudinal and transverse planes remain in the global sliding mode state, so:

$$\begin{aligned} S_1 &= \lambda_2 + \frac{n_1\lambda_1}{Y_f - Y} + Q_1 = 0 \\ S_2 &= \xi_2 - \frac{n_2(2Y + A)}{Y^2 + AY + B}\xi_1 + Q_2 = 0 \end{aligned} \tag{21}$$

The analytical solutions of λ_1, λ_2, ξ_1, ξ_2 can be obtained from the relationship between the above equation and the design state variables. In order to solve the problem conveniently, we take the $n_1 = n_2 = m_1 = m_2 = 3$ and it can be obtained that:

$$\begin{aligned} \lambda_1 &= (Y_f - Y)^3\left[\frac{\lambda_1(0)}{Y_f^3} + \left(\frac{\lambda_2(0)}{Y_f^3} + \frac{3\lambda_1(0)}{Y_f^4}\right)Y\right] \\ \lambda_2 &= (Y_f - Y)^2\left[-3\left(\frac{\lambda_1(0)}{Y_f^3} + \left(\frac{\lambda_2(0)}{Y_f^3} + \frac{3\lambda_1(0)}{Y_f^4}\right)Y\right) + \left(\frac{\lambda_2(0)}{Y_f^3} + \frac{3\lambda_1(0)}{Y_f^4}\right)(Y_f - Y)\right] \\ \xi_1 &= (Y^2 + AY + B)^3\left[\frac{\xi_1(0)}{B^3} + \left(\frac{\xi_2(0)}{B^3} - \frac{3A\xi_1(0)}{B^4}\right)Y\right] \\ \xi_2 &= (Y^2 + AY + B)^2\left[3(2Y + A)\left(\frac{\xi_1(0)}{B^3} + \left(\frac{\xi_2(0)}{B^3} - \frac{3A\xi_1(0)}{B^4}\right)Y\right)\right. \\ &\quad \left. + \left(\frac{\xi_2(0)}{B^3} - \frac{3A\xi_1(0)}{B^4}\right)(Y^2 + AY + B)\right] \end{aligned} \tag{22}$$

Further can be obtained:

$$\begin{aligned}
x &= \lambda_1 - \cot\gamma_f \cos\chi_f (Y - Y_f) + x_f \\
y &= \xi_1 + \cot\gamma_f \sin\chi_f (Y - Y_f) + y_f \\
\chi &= \mathrm{atan}(-(\xi_2 + \cot\gamma_f \sin\chi_f)/(\lambda_2 - \cot\gamma_f \cos\chi_f)) \\
\gamma &= \mathrm{atan}(\sin\chi / (\xi_2 + \cot\gamma_f \sin\chi_f))
\end{aligned} \tag{23}$$

Substitute (22) into (23) to get the analytical solution of x, y, χ,γ with respect to Y. Further derivative of the results with respect to Y can be obtained as follows:

$$\begin{aligned}
u_1 &= \lambda_2^{'} = -2(Y_f - Y)[-3(\frac{\lambda_1(0)}{Y_f^3} + (\frac{\lambda_2(0)}{Y_f^3} + \frac{3\lambda_1(0)}{Y_f^4})Y) \\
&+(\frac{\lambda_2(0)}{Y_f^3} + \frac{3\lambda_1(0)}{Y_f^4})(Y_f - Y)] - 4(Y_f - Y)^2(\frac{\lambda_2(0)}{Y_f^3} + \frac{3\lambda_1(0)}{Y_f^4}) \\
u_2 &= \xi_2^{'} = 2(Y^2 + AY + B)(2Y + A)[3(2Y + A)(\frac{\xi_1(0)}{B^3} + (\frac{\xi_2(0)}{B^3} - \frac{3A\xi_1(0)}{B^4})Y) \\
&+(\frac{\xi_2(0)}{B^3} - \frac{3A\xi_1(0)}{B^4})(Y^2 + AY + B)] \\
&+(Y^2 + AY + B)^2 (4(2Y + A)(\frac{\xi_2(0)}{B^3} - \frac{3A\xi_1(0)}{B^4}) \\
&+6(\frac{\xi_1(0)}{B^3} + (\frac{\xi_2(0)}{B^3} - \frac{3A\xi_1(0)}{B^4})Y))
\end{aligned} \tag{24}$$

By substituting the analytical solutions of (22), (23) and (24) into (17), it can obtain the analytical solution of $\chi^{'}$,$\gamma^{'}$ with respect to Y. Then substitute the result into Eq. (23) to get the expression of the lift coefficient. According to the lift coefficient, the value of drag coefficient can be deduced, so the value of drag can be obtained. The analytical solution of given velocity V with respect to Y is: $V^{'} = \frac{-D+mg\sin\gamma}{mV\sin\gamma}$, and the initial value of velocity V_0 is also known. Using these two known quantities and setting appropriate integral simulation step size, the velocity of the next moment can be obtained. In this way, the size of the terminal velocity V_f under A specific parameter can be iterated. Using this idea, the functional relationship between parameter A and the terminal velocity V_f can be written, and the expected value of parameter A under the terminal velocity can be solved by combining with the optimization algorithm (such as: fminsearch) in MATLAB.

4 Numerical Simulation and Analysis

This section mainly verifies the designed guidance law with impact angle and impact velocity. The initial position of the hypersonic vehicle in the reentry phase is set as $(x_0, y_0, z_0) = (0, 25000, 0)$ m, the target position is $(x_f, y_f, z_f) = (100000, 0, 50000)$ m, and the initial velocity is $V_0 = 2500$ m/s. The guidance parameters are $n_1 = n_2 = m_1 = m_2 = 3$, switching gains are $k_1 = k_2 = 0.0001$ and boundary layer thickness is $\zeta = 0.1$. The judgment basis of the end of simulation is that the distance of the projectile is less than or equal to 10 m or the attitude is less than or equal to 10 m.

4.1 Guidance Law Considering Only the Constraint of Impact Angle, i.e $A = 0$

Firstly, the initial track angle is set as: $\gamma_0 = -6°$, $\chi_0 = -8°$, and the expected impact angles (γ_f, χ_f) are respectively: $(-25°, -5°)$, $(-50°, 25°)$, $(-85°, 50°)$. The simulation results are as follows (Fig. 2):

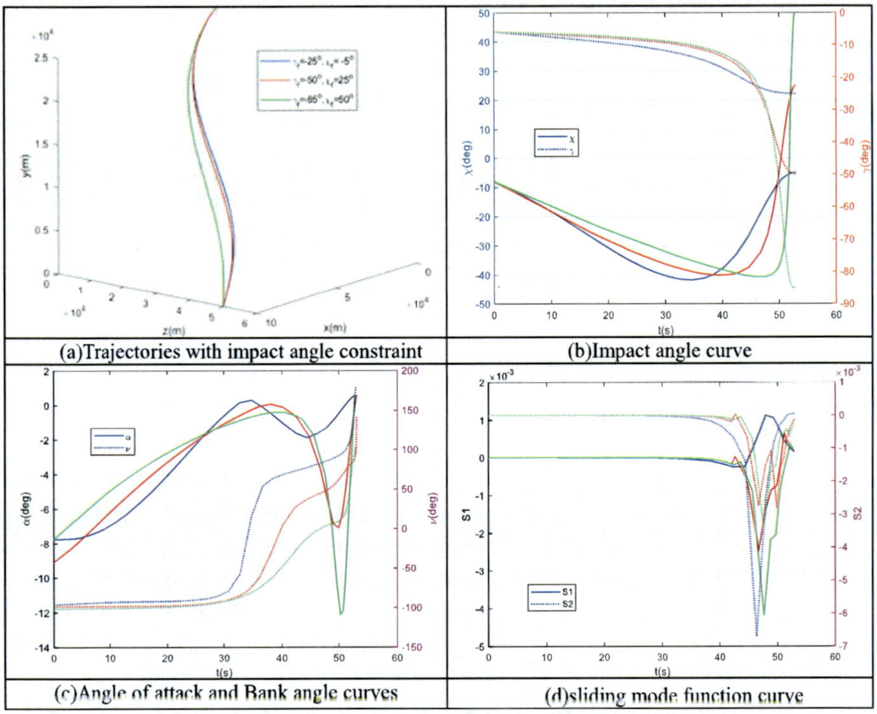

(a)Trajectories with impact angle constraint
(b)Impact angle curve
(c)Angle of attack and Bank angle curves
(d)sliding mode function curve

Fig. 2. Simulation results of constrained impact angle only

From the results, it can be found that under the guidance law, the vehicle can attack the desired target and the impact angle converges to the expected value. The larger the expected value of the impact angle is, the more curved the flight path will be, and the greater the amplitude of the corresponding guidance command will be. According to the curve of guidance command, The curves of attack angle and bank angle are smooth and continuous, which is beneficial to the tracking of attitude control system. In the later period of terminal guidance, it can be seen that the bank angle command will be between 90° and 180°, which indicates that the vehicle has carried out a large angle of rollover, which is conducive to attacking the target at positive angle of attack. It can be seen from the sliding mode function of the longitudinal and transverse planes that the global sliding mode can be realized, and its curve will produce a slight oversaturation command in the later stage, but it will also converge to zero at the end of the trajectory.

4.2 Constraints of Impact Angle and Impact Velocity are Considered Simultaneously

In this example, the initial values of the flight path angle and flight path azimuth angle are $\gamma_0 = -6°$, $\chi_0 = -8°$, and the expected impact angle is $\gamma_f = -85°$, $\chi_f = 0°$. Set the desired impact speed is $V_f = 1500\,\mathrm{m/s}$, $1400\,\mathrm{m/s}$. According to the above iterative idea, the parameter A is solved, and the corresponding A of the two cases are -5335.92 and -6332.11 respectively. The simulation results are as follows (Fig. 3):

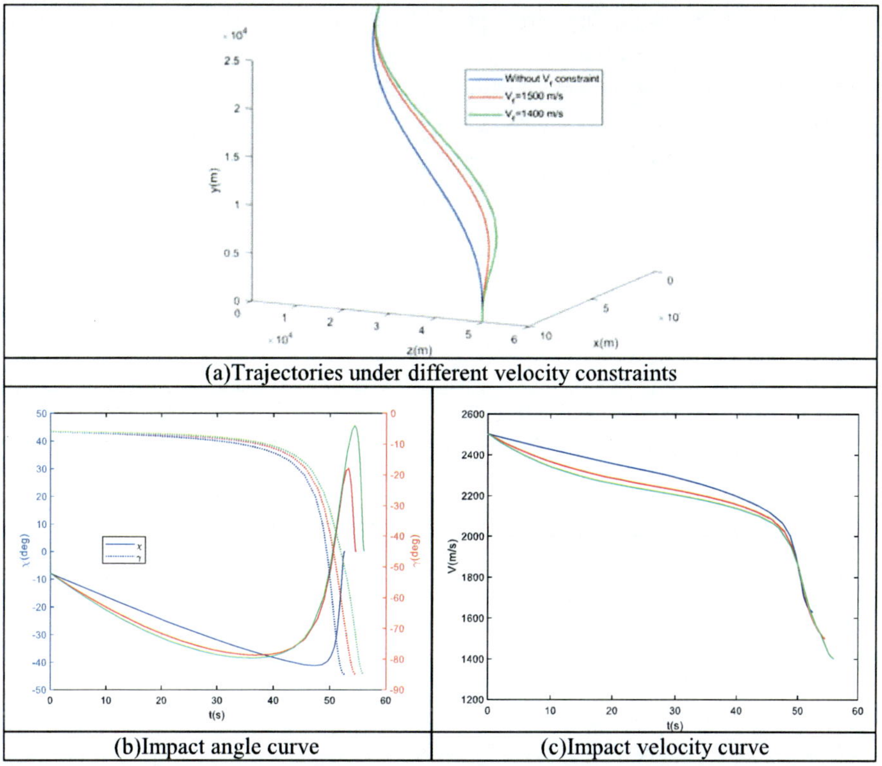

(a)Trajectories under different velocity constraints

(b)Impact angle curve

(c)Impact velocity curve

Fig. 3. Simulation results of simultaneously constraining impact angle and impact velocity

The above examples show that under the guidance law, the vehicle can realize the multiple constraints of terminal position, impact angle and impact velocity. From the longitudinal sliding mode control function and simulation results, it can be seen that the constraint of impact velocity mainly depend on transversal maneuvers. The essence of adjusting guidance parameter A is to change its transversal flight trajectory.

Without the velocity constraint, the impact velocity is 1629.49 m/s, while under the impact velocity constraint, the velocity is 1499.99 m/s and 1399.95 m/s respectively, which has a high control accuracy.

5 Conclusion

In this paper, a finite time sliding mode control method based on truncation function is proposed to solve the guidance problem of hypersonic vehicle under the constraints of impact angle and impact velocity. Based on the flight dynamics equation, the sliding mode function of the longitudinal and transverse planes are designed, and the global sliding mode is guaranteed by the auxiliary control parameters, the position and impact angle constraints of the guidance law are guaranteed. Through the analytical solution of the state quantities, the solution method of the guidance parameters in the transverse plane is derived, and the impact velocity is constrained by adjusting the transverse maneuver of the vehicle. Through simulation examples, firstly, it is verified that the guidance parameter $A = 0$ can meet the impact point accuracy and impact angle constraints; secondly, it is verified that different guidance parameters A can meet the impact speed constraints, and achieve the expected value. It can be seen that the guidance method proposed in this paper can meet the requirements of multi-constrained guidance, and has applicability and reliability.

Acknowledgements. This work is supported by Military scientific research program. And thank the senior and teachers for help.

References

1. Lu Pa. d, e, Doman D.B.b, e ,f, Schierman J.D.c, g (2006) Adaptive terminal guidance for hypervelocity impact in specified direction. J Guidance Contr Dyn 29(2): 269–278
2. Huifeng L, Yajie G, Zhaoying L (2013) Adaptive BTT terminal guidance law for hypervelocity vehicle. J Beijing Univ Aeronaut Astron 39(5):569–573
3. Zhaoshi D, Jiayuan S (2014) Continuous finite-time stabilization guidance law for terminal impact angle constrained flight trajectory. Yuhang Hsueh Pao (J Astron) 35(10):1141–1149
4. Zeng Y, Gu Y, Li J (2018) Terminal guidance law design with view angle and impact angle constraints associated with attack angle. Aerosp Contr 36(1):14–18
5. Du Z, Tang S (2013) Terminal guidance with impact angle constraint for hypervelocity reentry vehicle. Flight Mech 31(2):104–110
6. Ohlmeyer E.J.a, b, c and Phillips C.A.a, b, c (2006) Generalized vector explicit guidance. J Guidance Contr Dyn 29(2):261–268
7. Jianxue H, Kejun C, Hanyuan Z, Menglun Y (2007) Comparisons between reference-trajectory and predictor-corrector entry guidances for RLVs. J Natl Univ Defense Technol 29(1):26–34
8. Xie D, Wang Z, editors (2012) Adaptive proportional guidance law for reentry vehicles with impact angle and terminal velocity constraints. In: Proceedings of 10th World Congress on Intelligent Controland Automation; Beijing,China; 1 Natl Univ Def Technol, Coll Aerosp & Mat Engn, Changsha, Hunan, Peoples R China
9. Yu W, Chen W (2015) Trajectory-shaping guidance with final velocity and load factor constraints. ISA Trans 56:42–52
10. Kejun C, Luhua L and Yunhe M (2014) Launch vehicle flight dynamics and guidance. National Defense Industry Press, Beijing, pp 66–88
11. Xingjian L (2017) Introduction to missile overall design. National Defense Industry Press, Beijing, pp 2–8

12. Kuichen B, Changjian Z, Hong Z (2018) Gliding trajectory design of boost-gilde vehicle with terminal velocity restrict. Missiles Space Vehicl 4:6–10
13. Cheng W, Xugang W (2020) Terminal sliding mode control for hypersonic guided projectile. Syst Eng Electron 42(12):2859–2866

Check for updates

Dynamic Performance Test and System Identification of Air Rudder for Boost-Glide Aircraft

Xiaoshuai Fan(✉), Xibin Bai, Zhenyu Jiang, and Shifeng Zhang

College of Aerospace Science and Engineering, National University of Defense Technology, Changsha, China
s1156715841a@126.com

Abstract. Air rudder is the actuating mechanism of the boost-glide aircraft, which is used to control the flight attitude of the aircraft. Before the flight test, the time domain and frequency domain identification should be carried out by measuring the performance indexes of the air rudder, which is used for the 6-DOF simulation of the aircraft. This paper designs the test platform and test scheme of the air rudder of the boost glide aircraft, and tests the performance of the air rudder. The time domain and frequency domain identification methods are used to identify the four air rudder models of the boost-glide aircraft in this paper, and the verification scheme of cross-comparison between the time domain identification model and the frequency domain identification model is proposed. The test results show that the time domain identification method has low precision and can only represent the steady-state response of the air rudder. The frequency domain identification method has high accuracy, which is suitable for characterizing the change process of the air rudder from the initial response to the stable state, and the identification results can be applied well in the attitude control system of the boost-glide aircraft.

Keywords: Air rudder · Boost-glide aircraft · System Identification · Dynamic performance test

1 Introduction

Boost-glide aircraft is a hot research topic at present. By designing the angle of attack curve, the aircraft glides in the atmosphere and increases its range. Through the boost phase, glide phase, terminal guidance phase flight, it can accurately hit the target. It can effectively break through the intercept of the anti-aircraft system, strike the target, and has a strong battlefield adaptability by maneuvering in the atmosphere.

The air rudder system, as the actuator of the aircraft control system, can control the flight attitude of the aircraft and the flight trajectory of the aircraft. The computer on the aircraft sends out control instructions to the rudder system, which converts the received digital signals into PWM waves, and adjusts the deflection angle of the air rudder by adjusting the duty cycle of the waveforms. The air rudder rotates and drives the rudder blades to deflect, generating aerodynamic force and control the aircraft attitude [1]. The

H. Ding (ed.), *Aerospace Mechatronics and Control Technology*, Springer Aerospace Technology,
https://doi.org/10.1007/978-981-16-6640-7_3

execution efficiency and response characteristics of the rudder system directly determine the execution effect of the aircraft attitude control system. Thus, it determines the flight stability and strike accuracy of the aircraft.

The selection of aircraft air rudder requires comprehensive consideration of output torque, rotational speed and response characteristics. Because the interior of the air rudder is extremely complex, the design index and the actual index of the air rudder are difficult to match completely. And the performance of the air rudder and various indicators will change with the specific environment and time. Therefore, in order to ensure the stability and reliability of the aircraft actuator, the rudder system must be fully tested before used, so as to ensure the reliable flight of the aircraft and accurately hit the target. The actual transfer function and response curve of the air rudder are the core of the performance indexes of the air rudder, which must be tested and analyzed.

At present, the tests of the air rudder are mostly aimed at a specific air rudder, and the identification method and scheme platform used for identification are lack of generality, and are mostly used to identify the model aircraft air rudder. There is almost not identification for the air rudder system of the aircraft [2–6]. Model of aircraft air rudder generally has large deflection angle, small torque, good flying environment, it is easy to identify with high identification accuracy. The design of aircraft air rudder is complex, and it is faced with bad flying environment, so it needs to carry out large overload, large maneuver, and have large hinge moment, and require to response fast, and must deflect in a small limited angle range. Obviously, it has high requirements on the air rudder. There are some problems, such as difficulty in obtaining flight data, poor quality of test data and poor identification effect. At present, the identification of air rudder in literature is mostly limited to time domain identification or frequency domain identification, and new methods such as neural network [7] and subspace identification [8] are mostly adopted in the system identification. However, the identification results are not verified and are not widely applied in engineering with above algorithms.

To comprehensively consider the specific flight environment of aircraft, this paper designs a small boost-glide technology experimental aircraft (as shown in Fig. 1) and the general air rudder test platform (as shown in Figs. 2 and 3). The dynamic performance of the air rudder is planned to test, the time domain identification and frequency domain identification methods and results of contrast each other, determine the relatively reliable air rudder identification method and the recognition results. The designed air rudder identification scheme and platform have the advantages of high generality, convenient operation and low cost, which are suitable for large-scale popularization and application in the future.

2 Air Rudder Dynamic Performance Test Scheme and Test Platform Design

The dynamic performance test scheme and test platform of air rudder proposed in this paper are mainly used for boost-glide aircraft. The design of the boost-glide technology test aircraft as an example for testing. Its aerodynamic shape adopts a "$-\times$" layout. In order to improve the rudder efficiency, four air rudders are located at the tail of the test aircraft for boost glide technology. The overall physical figure is shown in Fig. 1.

Fig. 1. A physical view of the boost-glide technology test aircraft

2.1 Design of Dynamic Performance Test Platform for Air Rudder

The dynamic performance test platform of the air rudder is mainly to exert torque to the air rudder, to control the rotation of the air rudder, and measure the actual deflection angle of the air rudder through the angular displacement sensor, so as to continue to adjust the air rudder deflection, forming the closed-loop control of the air rudder deflection. In order to simulate the air rudder flight deflection more truly, the control instruction program is written by the onboard computer, and the control instruction is sent to the air rudder control system through the onboard computer, and the feedback instruction of the air rudder is collected. The work compared to a dedicated desktop machine to simulate the air rudder control instruction, using the onboard computer control air rudder to deflect directly closer to real flight environment, and can test the communication between onboard computer and air rudder, which is facilitate to realize program transplantation and replace the air rudder of tested. The designed test platform has high generality, air rudder test platform schematic diagram is as shown in Figure 3, the air rudder test platform for the material object is as shown in Fig. 2.

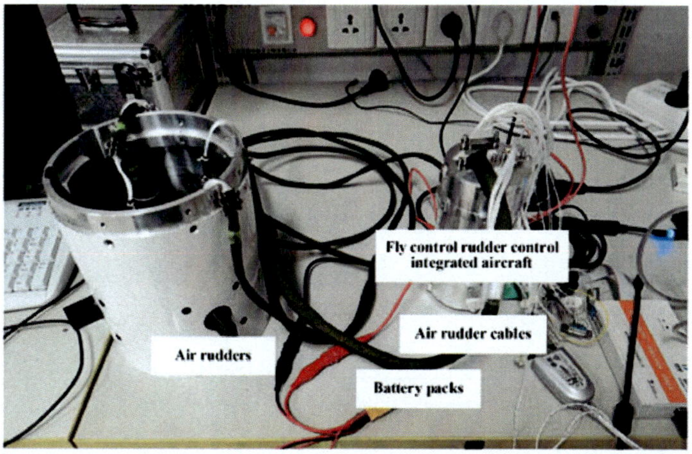

Fig. 2. Physical picture of dynamic performance test platform of air rudder

The flight control computer sends deflection instructions to the rudder control system through. The rudder control system converts the digital signals into PWM analog signals, then sends them to the air rudder for execution. The air rudder is mechanically connected with the rudder blade, and the air rudder drives the rudder blade to rotate. The rudder control system collects the actual air rudder deflection angle and feeds it back to the onboard computer to complete the air rudder deflection closed-loop control. The feedback of the air rudder is used as the input of identification, so the system identification of the air rudder model is carried out.

2.2 Design of Time Domain Dynamic Performance Test Scheme for Air Rudder

Through the time domain test of the air rudder, the transient performance index and steady performance of the air rudder can be analyzed, so as to identify the model of the system. Step signal, pulse signal, slope signal and parabola signal can be used as the input to test the time domain response. The rudder system in this paper is a position closed-loop servo system, and the step signal is added to test the dynamic performance of the air rudder in time domain [9].

When the Mach number is fixed, the lift coefficient of the boost-glide aircraft increases with the increase of the angle of attack, and when the angle of attack exceeds the critical value number, the aircraft will be in the state of loss speed. The lift coefficient not only does not increase, but also decreases sharply, and the aircraft is difficult to control. Therefore, for the engineering application, the aircraft's angle of attack$\alpha$is generally not more than 20°. For 1° rudder angle δ, the angle of attack generated by the aircraft body is the operational stability ratio to represent the air rudder operating efficiency, the general requirements of the operational stability ratio $|\alpha/\delta| = 1$, it require under extreme conditions $|\alpha/\delta| \in [0.5, 1.5]$. In order to ensure the flight safety and stability of the aircraft, the maneuverability can be large through aerodynamic design. Here, the maneuverability stability ratio under extreme conditions is 1. 5, then $20/1.5 = 13.3°$. Then the air rudder time domain test scheme takes the air rudder test instruction from 1°

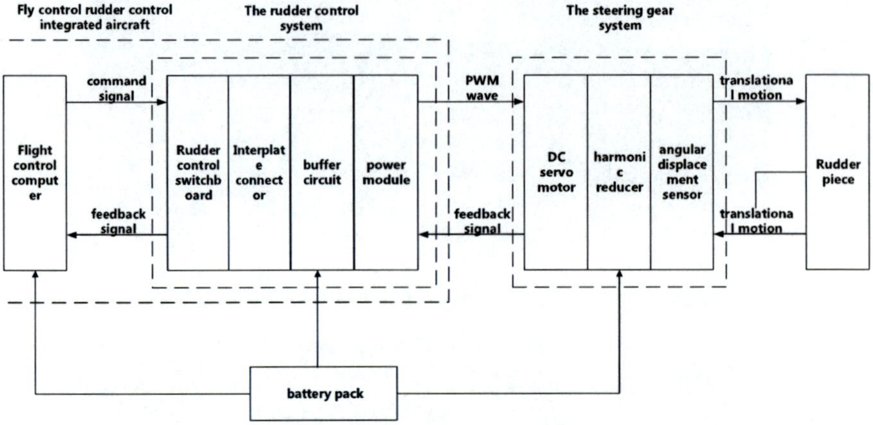

Fig. 3. Air rudder dynamic performance test platform schematic diagram

to 15°, steps for 1° with equispaced variation. Then the time domain performance of the air rudder can be fully measured, the derivation of the selected value of the test rudder deflection angle is shown in Fig. 4.

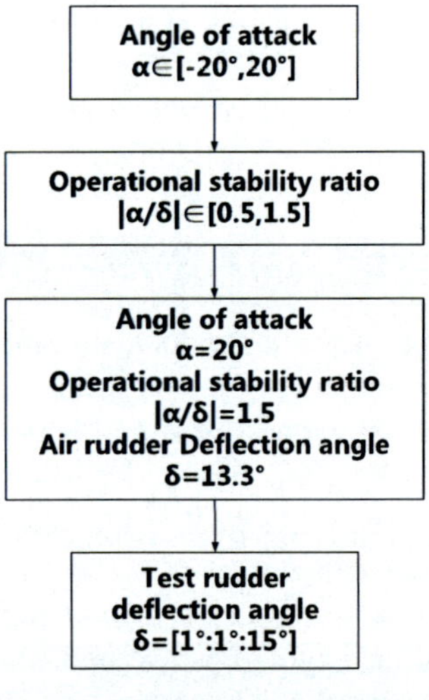

Fig. 4. Determine the value of the tested rudder deflection angle

2.3 Design of Dynamic Performance Test Scheme in Frequency Domain of Air Rudder

The frequency domain response method is a method to study the performance of the air rudder system by using the frequency performance of the air rudder. The model coefficient of the system can be identified by using the deterministic relationship between the ratio of stable output to stable input and the frequency of the air rudder. When the input signal is given, the stable-state output is the signal of the same frequency as the input signal, but the amplitude and phase are changed. Based on this, the open loop transfer function of the system can be obtained. Then the performance of the unit closed loop system can be analyzed by using the open loop transfer function of the system.

In the frequency domain scanning test of the air rudder, point-by-point sweep method, sinusoidal wave combination method, multi-harmonic difference trust sign method (SPHS) and linear frequency modulated pulse signal method (CHRIP signal) can be used [10]. The frequency sweep method is easy to operate, and the accuracy is high. It can fully exert the excitation to the air rudder, so the frequency sweep method is selected to test the frequency domain response of the air rudder.

The air rudder system of aircraft is a very complex opto-mechanical product, affected by various factors, there will be nonlinear links, if the test range beyond the linear work area, the identification accuracy will be reduced. The test instruction should be placed in the linear working area of the air rudder, and from Eqs. (1)–(8) should be used to select the test amplitude and frequency in the frequency domain of the air rudder.

With sinusoidal wave as the input signal, the test frequency f in the frequency domain depends on the maximum deflection angle rate of the air rudder, and the calculation formula is

$$2pfAS_c < \dot{\delta}_{\max} \tag{1}$$

In the above formula, f is the frequency to be selected for the frequency domain test of the air rudder, and A is the amplitude to be selected for the frequency domain test of the air rudder. In order to improve the test accuracy and process data, the test amplitude of the air rudder in the frequency domain is set as 1°.

Rated deflection rate of DC servo motor $\dot{\delta}_{\text{motor}}$ is

$$\delta_{\max}{=}58140\,°/\text{s} \tag{2}$$

Rated deflection rate of DC servo motor i is

$$i = 300 \tag{3}$$

Maximum angular velocity of air rudder $\dot{\delta}_{\max}$ is

$$\dot{\delta}_{\max} = \delta_{\max}/i = 193.8^{\circ}/s \tag{4}$$

The maximum frequency of the air rudder test, $f_{\max}$ can be calculated

$$f_{\max} = \dot{\delta}_{\max}/2\pi A = 30.85\text{Hz} \tag{5}$$

In order to ensure that the rudder is in the linear working area, take the safety factor S_c as

$$S_c = 10 \tag{6}$$

Then the maximum effective frequency of the air rudder test f is calculated

$$f = f_{\max}/S_{\text{c}}{=}30.57/10 = 3.57\text{ Hz} \tag{7}$$

To take the step size of test frequency as 0.2 Hz, then the test frequency f is

$$f = [0.2 : 0.2 : 3] \tag{8}$$

3 Time Domain and Frequency Domain System Identification Method of Air Rudder

3.1 Method for Identification of Air Rudder Time Domain System

The air-rudder system of the boost-glide aircraft can generally be approximated by the first-order system, second-order system or third-order system. With the increase of the order, the accuracy of the system will be improved, but the stability will be reduced. In actual systems, there are few systems of third-order or above. In this paper, the servo response is approximated as the following second-order system [11].

$$G(s) = \frac{\omega_n^2}{s^2 + 2\xi\omega_n s + \omega_n^2} \tag{9}$$

For the model parameter identification problem, according to the system response equation of the unit step signal applied to the second-order system, we can know.

If $0 < \xi < 1$, it's called underdamping system

$$x_0(t) = 1 - \frac{e^{-\xi\omega_n t}}{\sqrt{1-\xi^2}} \sin(\omega_d t + \theta) \tag{10}$$

Which are in the above formula: $\omega_d = \omega_n\sqrt{1-\xi^2}$, $\theta = \tan^{-1}\frac{\sqrt{1-\xi^2}}{\xi}$.

To let step response curve are normalized, the step response data at each time point is divided by the step signal amplitude, the step response curve is drawn. The peak time t_p and the maximum overshoot M_p are gotten from the figure. The second-order model parameters ξ, ω_n are obtained according to the following equation:

$$\begin{cases} \omega_n\sqrt{1 \quad \xi^2} = \dfrac{\pi}{t_p} \\ \dfrac{\xi}{\sqrt{1-\xi^2}} = -\dfrac{\ln(M_p)}{\pi} \end{cases} \tag{11}$$

Thus, the transfer function of the desired object can be obtained.

For critically damped or overdamped cases, that is $\xi \geq 1$, only T and ξ can determine the transfer function of the system.

G(s) can be broken down into:

$$G(s) = \frac{1}{T^2(s + \omega_1)(s + \omega_2)} \tag{12}$$

In the formula above formulate: $\omega_1 = \frac{1}{T}\left[\xi + \sqrt{\xi^2 - 1}\right]$, $\omega_2 = \frac{1}{T}\left[\xi - \sqrt{\xi^2 - 1}\right]\omega_1, \omega_2$ are all real numbers and they're all greater than zero,$T = \frac{1}{\sqrt{\omega_1\omega_2}}$, $\xi = \frac{\omega_1 + \omega_2}{\sqrt{\omega_1\omega_2}}$.

The transfer function can be reduced to

$$G(s) = \frac{\omega_1 \omega_2}{(s + \omega_1)(s + \omega_2)} \tag{13}$$

Therefore, identification of the transfer function is transformed into a solution ω_1 and ω_2.

When the input signal is the unit step function, the inverse Laplace transform of the above equation is applied to obtain the unit step response of the system in the time domain

$$y(t) = 1 - \frac{\omega_2}{\omega_2 - \omega_1} e^{-\omega_1 t} + \frac{\omega_1}{\omega_2 - \omega_1} e^{-\omega_2 t} \tag{14}$$

Then

$$1 - y(t) = \frac{\omega_2}{\omega_2 - \omega_1} e^{-\omega_1 t} - \frac{\omega_1}{\omega_2 - \omega_1} e^{-\omega_2 t} \tag{15}$$

Because of its exponential characteristics, the curve fitting can be realized directly through the nonlinear least square algorithm. Then ω_1 and ω_2 can be calculated, the transfer function of the system can be obtained.

3.2 Identification Method of Frequency Domain System of Air Rudder

The frequency sweep method is adopted to select a series of frequency test points, and the constant amplitude sinusoidal signal is input at each frequency point as the excitation of the tested system, and the output signal is sampled to obtain the response of the system. The amplitude gain and phase difference of the rudder system at different frequencies can be obtained by means of fitting analysis and correlation analysis, and then the coefficients of the rudder system model can be identified by Levy method. In this paper, the correlation analysis method is used to solve the amplitude phase of the system.

The air rudder control instruction $x(t)$ and feedback voltage signal $y(t)$ expressed as

$$\begin{cases} x(t) = A \sin(\omega t + \theta_1) + N_x(t) \\ y(t) = B \sin(\omega t + \theta_2) + N_y(t) \end{cases} \tag{16}$$

In the above formula: $N_x(t)$, $N_y(t)$ are the noise.

The autocorrelation functions of the two signals are solved by comparing the identification results and the experimental results

$$\begin{cases} R_x = \dfrac{1}{T} \displaystyle\int_0^T x(t)x(t + \tau)dt \\ R_y = \dfrac{1}{T} \displaystyle\int_0^T y(t)y(t + \tau)dt \end{cases} \tag{17}$$

If $\tau = 0$.

The amplitude of the air rudder system at this frequency is

$$A_m = \frac{B}{A} = \sqrt{\frac{R_y(0)}{R_x(0)}} \tag{18}$$

And it's a cross-correlation function of two signals

$$\begin{aligned} R_{xy}(\tau) &= \frac{1}{T}\int_0^T x(t)y(t+\tau)dt \\ &= \frac{1}{T}\int_0^T [A\sin(\omega t + \theta_1) + N_x(t)] \\ &* \left[B\sin(\omega(t+\tau) + \theta_2) + N_y(t+\tau)\right]dt \end{aligned} \tag{19}$$

If $\tau = 0$, it can be concluded that the phase of the air rudder system at this frequency is

$$\varphi = \theta_2 - \theta_1 = -\arccos\frac{2}{AB}R_{xy}(0) \tag{20}$$

Based on the test data of different frequencies, the amplitude and phase of the air rudder system are solved through correlation analysis, and the response data in frequency domain is obtained.

Assume that the transfer function of the air rudder system under test is

$$G(s) = \frac{b_0 + b_1 s + \cdots + b_m s^m}{1 + a_1 s + \cdots + a_n s^n} \quad n \geq m \tag{21}$$

In the above formula, $s = j\omega$, to make it express easier, it is unified into ω. Then the numerator and denominator of the system transfer function are written as functions of ω, they are $N(\omega)$, $D(\omega)$. Then the transfer function $G(s) = \frac{N(\omega)}{D(\omega)}$. The real and imaginary parts of the transfer function are also written as functions of ω, they are $\mathrm{Re}(\omega)$, $\mathrm{Im}(\omega)$. The real and imaginary parts of transfer function calculated by the measured data are also written as functions of ω, they are $\overline{Re}(\omega)$, $\overline{I}m(\omega)$.

The test data in the experiment are

$$\Phi(\omega) = \overline{Re}(\omega) + \mathrm{j}\overline{I}m(\omega) + \varepsilon \tag{22}$$

In the above formula, $\mathrm{Re}(\omega) = \overline{A}_m\cos\varphi$, $\mathrm{Im}(\omega) = \overline{A}_m\sin\varphi$.

The error is expressed as ε, ε is white noise, $E(\boldsymbol{\varepsilon}) = 0$. In the frequency ω, the error between the estimated frequency response and the measured frequency response is

$$\begin{aligned} E(\omega) &= \Phi(\omega) - H(\omega) \\ &= (\overline{Re}(\omega) + j\overline{I}m(\omega) - \varepsilon) - \frac{N(\omega)}{D(\omega)} \end{aligned} \tag{23}$$

Let's take the index function

$$J(\omega) = \sum_{i=1}^{L} \|E(\omega)\|^2 \tag{24}$$

This is a solution to the minimum value problem, the optimization algorithm can be obtained by polynomial coefficient. Then, the transfer function of the system can be obtained. Here, it can make the system residual minimum by using the least square method, then the transfer function of the system can be obtained.

Here, the air rudder system is approximately described by a second-order system, namely:

$$G(s) = \frac{b_0}{1 + a_1 s + a_2 s^2} \tag{25}$$

Then the transfer function of the system can be expressed as

$$\begin{bmatrix} \mathrm{Re}(\omega) \\ \mathrm{Im}(\omega) \end{bmatrix} = \begin{bmatrix} 1 & \mathrm{Im}(\omega)\omega & \mathrm{Re}(\omega)\omega^2 \\ 0 & -\mathrm{Re}(\omega)\omega & \mathrm{Im}(\omega)\omega^2 \end{bmatrix} \begin{bmatrix} b_0 \\ a_1 \\ a_2 \end{bmatrix} \tag{26}$$

According to the least square principle, the above equation is simplified as

$$\mathbf{Y} = \mathbf{HX} + \boldsymbol{\varepsilon} \tag{27}$$

In the above formulate, $\mathbf{Y}$ is the observation vector, it is a column vector composed of the real part and imaginary part of the system transfer function solved by experimental data. $\mathbf{H}$ is the coefficient matrix, $\mathbf{X}$ is the parameter vector identified, $\boldsymbol{\varepsilon}$ is the error vector. Assuming that $E(\boldsymbol{\varepsilon}) = 0$.

Then, $\overset{\wedge}{\mathbf{X}}$ is the least square estimation.

$$\hat{\mathbf{X}} = \begin{bmatrix} \hat{b}_0 & \cdots & \hat{b}_m & \hat{a}_1 & \cdots & \hat{a}_n \end{bmatrix}^T = \left(\mathbf{H}^T\mathbf{H}\right)^{-1}\mathbf{H}^T\mathbf{Y} \tag{28}$$

Then, the transfer function of the air rudder system obtained through the least square identification is

$$\hat{G}(s) = \frac{\hat{b}_0 + \hat{b}_1 s + \cdots + \hat{b}_m s^m}{1 + \hat{a}_1 s + \cdots + \hat{a}_n s^n} \quad n \geq m \tag{29}$$

The model of the system can be identified by Eq. (29).

4 Test Results and Analysis of Dynamic Performance of Air Rudder

4.1 Identification Method of Frequency Domain System of Air Rudder

Accomplished the identification of the air rudder transfer function through the following seven steps:

(1) Contrast air rudder instructions and feedback;
(2) Align instruction and feedback with start time. Align the starting point of feedback data of air rudder with different deflection angles.

(3) The step response data is normalized;
(4) Average step response data;
(5) Calculation of step response characteristics;
The maximum point of the averaged step response curve is $y_{\max}$. The time corresponding to the maximum point is the peak time t_p, the overshoot can be calculated according to the following formula:

$$\sigma = \frac{y_{\max} - y_{\infty}}{y_{\infty}} \times 100\% \tag{30}$$

In the formula above, y_{∞} is the stable-state value of step response. The mean value of data from 1 to 2 s is expressed as: $y_{\infty}=\overline{y}_{t=1\sim2}$.
(6) Air rudder transfer function calculation.
The damping coefficient is calculated according to Eq. (10). Then ζ and the natural frequency ω_n can be calculate, the transfer function of four air rudders are obtained.
$G_1(s)$, $G_2(s)$, $G_3(s)$, $G_4(s)$ are the transfer functions of four air rudders respectively, as shown in Eq. (31).

$$\begin{aligned} G_1(s) &= \frac{639.2}{s^2 + 44.42s + 639.2} \\ G_2(s) &= \frac{725.1}{s^2 + 48.13s + 725.1} \\ G_3(s) &= \frac{457.6}{s^2 + 38.59s + 457.6} \\ G_4(s) &= \frac{527.9}{s^2 + 40.1s + 527.9} \end{aligned} \tag{31}$$

(7) The identification results and experimental results are compared and analyzed.
The results of the rudder 1 are analyzed as the example, the analysis of the other three steering machines can be obtained in the same way.
Figure 5 shows the comparison between the experimental step response curve and the identified step response curve of the air rudder 1. The x-coordinate is the time, and the unit is seconds. Then the rudder starts responding to the deflection instruction, current time is repressed as original point. The y-ordinate is the feedback value of the deflection angle collected by the air rudder, and the unit is degree. 'experiment' is the test step response curve, and 'matching' is the identified step response curve. As can be seen from Fig. 5, the fitting effect of the stable amplitude is good, which can accurately represent the characteristics of the air rudder system. However, the fitting results of the response process are not completely coincident with the experimental results, the results of identification and actual results are slightly different a lot. In general, the time domain identification effect is not very ideal, the identification model is not very accurate. Therefore, the frequency domain identification method is considered.

4.2 Test Results and Analysis of Dynamic Performance of Air Rudder in Frequency Domain

The amplitude and phase of the rudder system are solved by the correlation analysis method, then the coefficients of the rudder system model are identified by Levy method.

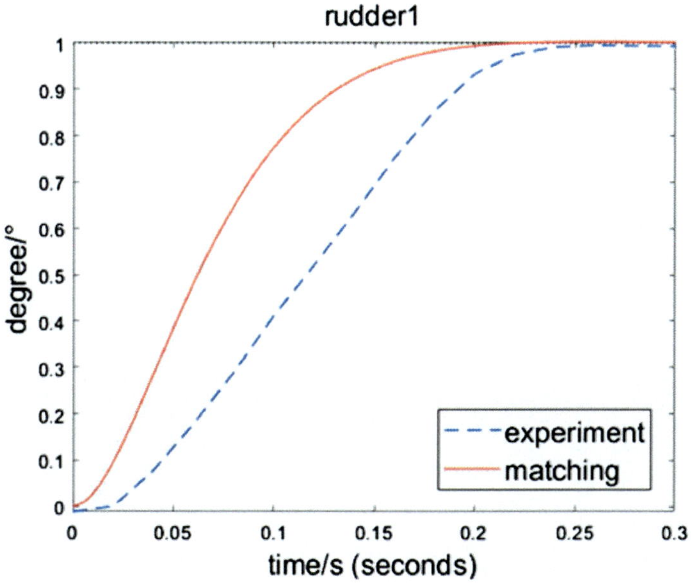

Fig. 5. Comparison of experimental results and identification results of air rudder 1

The test input is a sinusoidal signal with amplitude of 1°, the frequency varies from 0.2 Hz to 3 Hz, step size of 0.2 Hz and initial phase of 0°. For the experimental data obtained, the correlation analysis method is used to solve the amplitude and phase of the system. Then Levy method is used to carry out the system identification. The transfer function of identification is shown in the Eq. (32)

$$\begin{aligned} G_1(\mathrm{s}) &= \frac{527.1}{s^2 + 27.52s + 611.5} \\ G_2(\mathrm{s}) &= \frac{403.5}{s^2 + 27.32s + 456.9} \\ G_3(\mathrm{s}) &= \frac{367.7}{s^2 + 27.07s + 410.7} \\ G_4(\mathrm{s}) &= \frac{558.5}{s^2 + 30.61s + 604.6} \end{aligned} \tag{32}$$

As is shown in Figs. 6 and 7, 'experiment' represents the experimental value and 'matching' represents the identification value. By analyzing the amplitude-frequency characteristic curve and phase-frequency characteristic curve, the identification results are close to the experimental results, which indicates that the identification results are relatively reliable.

4.3 Mutual Verification Analysis of Time Domain and Frequency Domain Identification Results

The identification results in frequency domain are put into the step response test data in time domain to verify whether the identification results are consistent with the measured

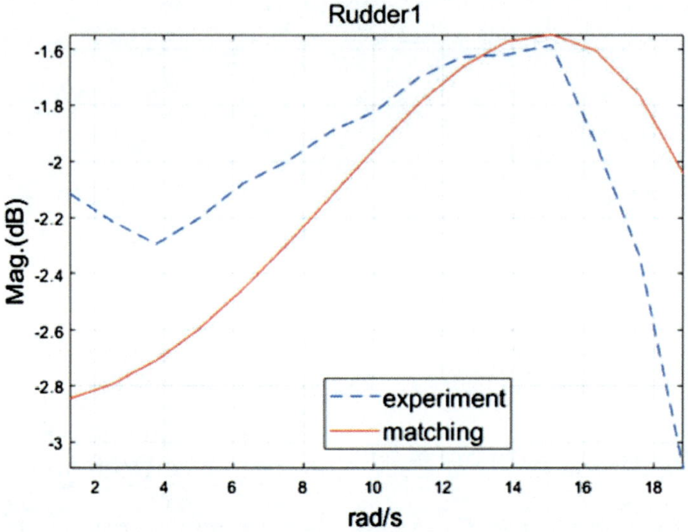

Fig. 6. Amplitude frequency characteristic curve of air rudder 1

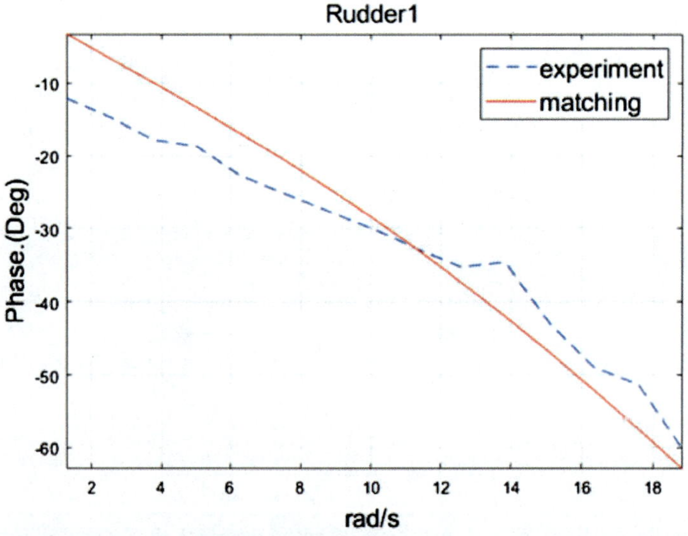

Fig. 7. Phase frequency characteristic curve of a air rudder 1

data. The time domain identification results are put into the frequency domain test data to verify whether the identification results are consistent with the measured data. The identification results can be verified by mutual analysis.

The results of the rudder 1 are analyzed as example, the analysis of the other three steering machines can be obtained in the same way.

(1) Put the frequency domain identification results (Eq. (32)) into the time-domain step response experimental data, and the results are shown in Fig. 8. The x-ordinate represents time, the unit is seconds. The y-ordinate is the amplitude of step response, and the unit is degree. 'experiment' represents the step response curve consists of experimental data. 'relevant' represents the step response curve of the identification results based on correlation analysis.

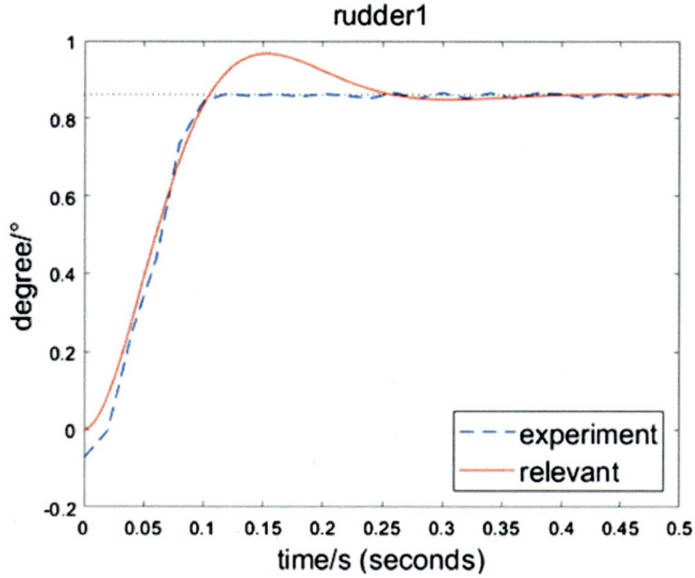

Fig. 8 Comparison of frequency domain identification results and time domain step response experimental values of air rudder 1

As can be seen from Fig. 8, the frequency domain identification results are highly coincident with the step response curves measured by experiments, which indicates that the frequency domain identification method is feasible and reliable, and the identification results have been well verified.

(2) Put time domain identification results (Eq. (31)) into the experimental data curve of amplitude-frequency characteristics and phase-frequency characteristics experimental data in the frequency domain, and the results are shown in Figs. 9 and 10. In the figure, the x-coordinate represents frequency and the unit is rad/s, and the y-coordinate represents amplitude gain and the unit is dB in the amplitude-frequency characteristic curve. The y-ordinate represents phase and the unit is degree in the phase-frequency characteristic curve. 'experiment' represents the amplitude-frequency characteristic curve and phase-frequency characteristic curve consists of experimental data, and 'time domain' represents the amplitude-frequency characteristic curve and phase-frequency characteristic curve generated by time-domain identification results.

As can be seen from Figs. 9 and 10, the phase-frequency characteristic curve of the identification results and the experimental data are in good agreement, while

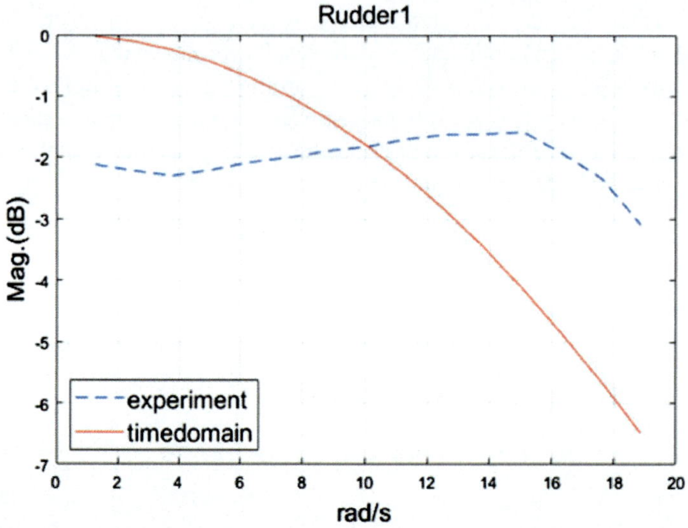

Fig. 9. The time domain identification results of air rudder 1 are compared with the experimental amplitude-frequency characteristic curves in frequency domain

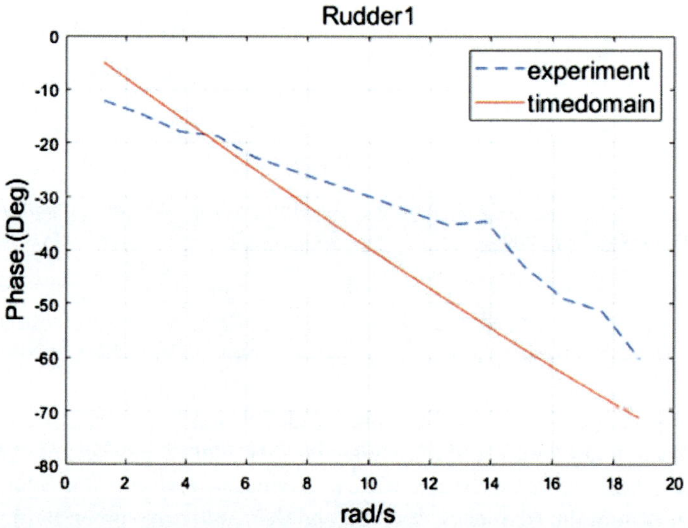

Fig. 10. The time domain identification results of air rudder 1 are compared with the experimental phase frequency characteristic curves in frequency domain

the amplitude-frequency characteristic curves are in poor agreement. The variation trend of time–amplitude-frequency characteristic curve between the identification results and the experimental results is relatively close at high frequency, while the variation trend is quite different at low frequency, which indicates that the high frequency response characteristics and low frequency response characteristics of the rudder are different. This phenomenon is determined by the physical characteristics of the air rudder itself. The time domain identification method and frequency domain identification method adopted to the air rudder models are inconsistent, which results in different identification results. From the identification results, the frequency domain identification result is more ideal.

4.4 Engineering Application of the Identification Results of Air Rudder

The identified transfer function can be used in the air rudder transfer function module of the control law verification model in Fig. 11, namely the air rudder transmission function $G(s)$. When applying the identified transfer function of the air rudder into attitude control system, the model of the whole control system can be established more accurately, which plays a key role in the maneuvering and precision guidance of the boost-glide aircraft in the atmosphere.

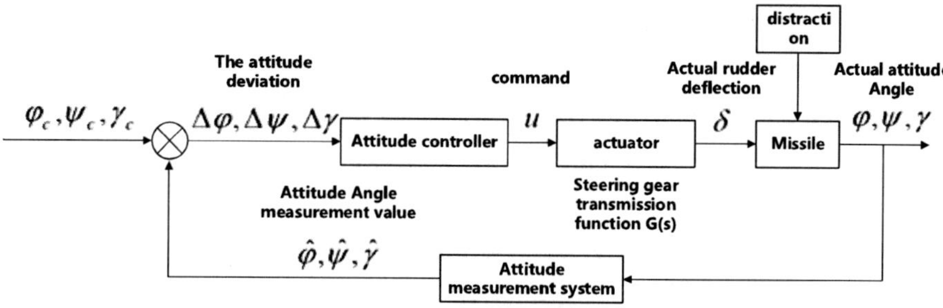

Fig. 11. The application of air rudder identification in attitude control system

5 Conclusion

A test platform for the dynamic performance of the air rudder of the boost-glide aircraft is designed in this paper, which has the characteristics of high integration, convenient operation, strong versatility, etc. It realizes the lightweight design. By using the aircraft onboard computer and air rudder in actual flight to test the air rudder, which close to the real working state of the aircraft in actual flight. It can not only test the performance of the air rudder effectively, but also test the communication link between the aircraft onboard computer and air rudder, which gives full attention to the core control function of the aircraft onboard computer.

According to the working state and engineering experience of the air rudder during the actual flight of the boost-glide aircraft, the test scheme of the air rudder in time domain and frequency domain is determined. The test scheme in time domain and frequency domain of the air rudder selected has strong applicability. It can flexibly modify the test parameters according to the performance of the air rudder, and has high versatility.

The time domain dynamic performance test and system identification method of the air rudder based on the step response and the frequency dynamic performance test and system identification method of the air rudder based on the correlation analysis method are used to complete the performance test and system identification of the air rudder system, and the two methods verify each other. The test results show that the air rudder works normally. Both identification methods are real and reliable, and the frequency domain identification method is more complicated, but the accuracy is higher. The time domain identification method is easy to operate, but the identification accuracy is lower. The identification method can be determined according to the actual situation and the accuracy requirement.

The dynamic performance test and system identification results of the air rudder can be used to the attitude control system of the boost-glide aircraft and control the air rudder to complete the specified action. With the air rudders, the boost-glide aircraft can maneuver in the atmosphere to increasing penetration and deliver precise strikes on targets. Therefore, the designed air rudder test platform and test scheme have broad application prospects in the future.

Acknowledgments. Thanks to my tutor's advice, I finish this survey with the help of my teachers ,thanks them very much. I would like to pay high tribute to National University of Defense Technology for supporting the platform to carry out this research.

References

1. Yang C (2016) Princple of aircraft aeroelasticity, 3rd edn. Beihang University Press, Bei-jing, pp 203–208
2. Chen H, Zhuang H, Ma S et al (2014) Model identificationof electromechanical actuator based on tests. Aeronaut Comput Tech 44(4):6–9
3. Ma M, Zhou S, Wang B et al (2016) Parameter identificationof electric actuator servo system on aircraft. Comput Simul 33(2):88–92
4. Zhu M, Cao G, Zhang Z et al (2011) The system identification of pneumatic actuator based on levy method. J Projectiles, Rockets, Missiles Guidance 31(6):69–72
5. Zhang J, Liu H, Jia H et al (2008) Model identification andcorrector design for servo system of electromechanical actuator. Opt Precision Eng 16(10):1971–1976
6. Gao S (2006) Research and experiment of system identificationbased on real-gained data. Harbin Institute ofTechnology, Harbin, pp 25–38
7. Zhihong W, Ruifeng Y, Chenxia G, Shuangchao G (2020) Identification method of air rudder system based on improved adaptive genetic algorithm. Sci Technol Eng 20(11):4436-4441
8. Yukai S, Renjia Z, Zhigang W, Chao Y, Yang Y (2020) Dynamic characteristics test and system identification of model aircraft air rudder. J Beijing Univ Aeronaut Astronsut 46(**02**):294–303
9. Qinghua Z, Weihua Z (2011) Unmanned flight control system experiment tutorial. National Defense Industry Press, Beijing
10. Yinhui Z (2011) Hardware-in-the-loop simulation of suborbital vehicle control system based on DSPACE. Zhang Weihua. National University of Defense Technology, Tutor
11. Li T, Zhang S, Yang H, Zhang Y (2014) Design and verification of control system for small solid rocket based on extended state observer. J Solid Rocket Technol 37(06):749–755

DSMC Study for Effects of Angles of Attack on Closed Cavity of Space Vehicle in Hypersonic Rarefied Flow

ShaoJin Xiang(✉) and ShuZhou Fang

School of Aerospace and Engineering, Beijing Institute of Technology, Beijing 10081, China
shaojinxiang@foxmail.com

Abstract. In order to study the flow characteristics and the heat flux of cavity with length-to-depth ratio of 6 on the surfaces of the hypersonic reentry aerospace vehicles in rarefied gas flow, Direct Simulation Monte Carlo method (DSMC) with the adaptive grid is used. The free stream at a Mach Number of 8, an altitude of 60 km, and the angles of attack (AOA) ranging from 0° to 60°.The results show that the closed cavity changes its type when AOA changes and is back to closed type when AOA is 60°. Increasing AOA sometimes does not help free-stream to get into the cavity because the stream strikes the plate on the front edge of the cavity, and form a shock wave that changes the direction of the free-stream flow. The low-speed and high pressure region inside the cavity extends to the upper zone of the cavity due to the shock wave, the speed is getting smaller and the pressure is getting higher. When AOA is 40° and 50°, the heat flux of three surfaces of the cavity is higher than other situation.

Keywords: Hypersonic rarefied flow · Direct simulation Monte Carlo method · Cavity · Angle of attack · Shock wave

1 Introduction

As for the near-space vehicles, including re-entry vehicles, cruise vehicles, and interstellar exploration vehicles, aerodynamic heat and forces in rarefied gas flow have an important impact on the design of their structure and thermal protection system. Standing on the point of design and research, it is usually assumed the surface is smooth. However, in practical applications, it is difficult to avoid the appearance of various shapes of cavities on the surfaces of space vehicle, such as the gaps between the heat shields [1], the hatch and covering cap [2], and some gaps caused by sensor installation. When the hypersonic flow passes through these structures, the boundary layer may separate at the entrance. The interference of the cavity will increase the turbulence and accelerate the transition of the boundary layer. The vortex will be rolled up inside and a lot of heat will be drawn into the cavity. The radiation heat dissipation effect will be blocked because of the narrow cavity [3], etc. Therefore, the study of the cavities about their flow field, thermal environment and internal vortex is very necessary.

H. Ding (ed.), *Aerospace Mechatronics and Control Technology*, Springer Aerospace Technology,
https://doi.org/10.1007/978-981-16-6640-7_4

Charwat et al. [4, 5] studied the flow around cavities with different length-to-depth ratios (the ratio of the length L of the cavity to the depth D), and divided them into three categories according to the flow field structure: open, closed and transition style. Open cavity flow ($1 < L/D < 10$), the external flow directly crosses the cavity, and the shear layer separated from the front wall is attached to the top of the back wall, forming a vortex inside the cavity. Closed cavity flow ($L/D > 14$), the shear layer adheres to the bottom of the cavity, and then separates again at the intersection of the back wall and the freestream, inside the cavity there are two vortexes. The transitional cavity flow ($10 < L/D < 14$) is a state between the above two flows. In the 1970s, the Langley Research Center carried out a series of experiments [6, 7] to study the thermal environment of the cavity. Tong [8] investigated the heat shields by using N-S equation, he simplified the two-dimensional gap flow to two linear superposition of two one-dimensional flows. Chun et al. [9] investigated the temperature, pressure and heat transfer of the seal structure with gap and cavity by using fluid–structure method, and the results indicated the structural layout of the seal plays a decisive role in thermal conditions. Qiu et al. [10] investigated the flow characteristics and thermal environment of gap, they indicated the heat flux distribution affected by the vortexes inside the gap. Gong et al. [11] and Tang [12] experimented with a flat plate model in a shock wave wind tunnel to measure heat and studied the influence of angles of attack, width, depth, and step height on the thermal environment. Experimental data reveal the peak heating most likely to appear on the top edge of gap.

Above work is carried out in continuum regime, and there are few studies about cavity in rarefied flow. Compared with the free molecular flow close to the vacuum, the transition flow is more complicated due to the rarefied gas effect and non-equilibrium effect etc. Santos [13–16] investigated the heat transfer coefficient, pressure coefficient, and surface friction coefficient of various shapes of cavities in rarefied flow, such as forward and backward steps, cavities with varying length-to-depth ratios, etc. Guo [17] et al. investigated the flow field of a cavity with length-to-depth ratio 1–8, he indicated the cavity turn to closed type when L/D is 6. Jin et al. [18] investigated the effects of rarefied gas and three-dimensional property on flow-field structure inside the cavity and heat flux over the surfaces, the results indicated the vortex became slender and the heat flux concentrated to the top region of the downstream surface of the cavity. Zhang et al. [19] investigated the flow characteristics of flow field in the cavity by using DSMC method to simulate different height and the Mach number conditions. The results indicated the flow structure in the cavity would change in type at specific aspect.

During the re-entry process, the aircraft not only needs to cross different altitudes, but also needs to change the angle of attack. There is few research concerning about the AOA. Based on the work of Guo [17], this paper uses the DSMC method to study the changes in the internal flow field structure and thermal environment of the closed cavity under different angles of attack situation.

2 Computational Method and Example Verification

2.1 DSMC Introduction

Bird [20] put forward the concept of Direct Simulation Monte Carlo (DSMC) in 1966, and verified its feasibility in hypersonic transition flow. The DSMC method is a direct simulation method based on kinetic theory of molecules [22], using simulated molecules to represent a large number of real gas molecules. During the simulation process, the computer stores the position, velocity and internal energy of the simulated molecule in the memory. The calculation time step is smaller than the physical collision time to decouple the movement of the molecule and the collision process. Then a certain number of collision pairs are selected by statistical methods, and use the collision model to calculate. Computers update the data of the simulated molecule. After enough time steps to make the statistical error small enough and ensure the stability of the flow field, properly sampling the microscopic properties of the simulated molecule to obtain the gas macroscopic characteristics, such as velocity, temperature, density, shear stress, pressure, etc.

2.2 Validation of the DSMC Code

The DSMC calculation program in this paper adopts the SPARTA (Stochastic Parallel Rarefied-gas Time-Accurate Analyzer) program from Sandia National Laboratory in the United States. This program uses adaptive structural grids. The molecular collision uses variable hard sphere (VHS) model, and no-time-counter (NTC) method for sampling. Energy exchange uses Larsen-Borgnakke model. The wall is set to diffuse reflection and the temperature is constant.

In order to verify the ability of the SPARTA program to capture complex physical phenomena and the accuracy of numerical simulation, the following will calculate the scaled model of the Mars Pathfinder (Mars Pathfinder) exploration vehicle—70° blunt cone vehicle model. Allegre et al. [21] selected this model to conduct rarefied flow experiments in multiple independent wind tunnels, and gave a large amount of heat transfer experimental data that could be used for verification. The corresponding calculation conditions are shown in Table 1. Reynolds number *Re* and the gas rarefaction parameter $\left(\overline{V} = Mach\big/(\mathrm{Re})^{0.5}\right)$ is calculated based on the blunt cone head diameter of 50 mm.

Table 1. Calculation conditions for blunt cone vehicle

Case	A	B
Gas	N_2	N_2
Velocity	20.2 Mach	20 Mach
Temperature	1100 K	1100 K
Pressure	3.5 Bar	10 Bar
Re	1420	4175
$\overline{V}$	0.53	0.31

The schematic diagram of blunt cone is shown in Fig. 1a. The distribution of test points in the low-density SR3 wind tunnel experiment is shown in Fig. 1b. The heat is a function of the distance S from the test point to the leading edge stagnation point. S/Rn is a dimensionless parameter about this distance. In the calculation, a structural grid based on the Cartesian coordinate system is used, the first-level grid is divided into 800×800, and the second-level grid is divided by local adaptive technology according to the mean free path of the particles. The freestream comes from the left boundary and its direction is parallel to the rotation axis. Other boundaries set free interface.

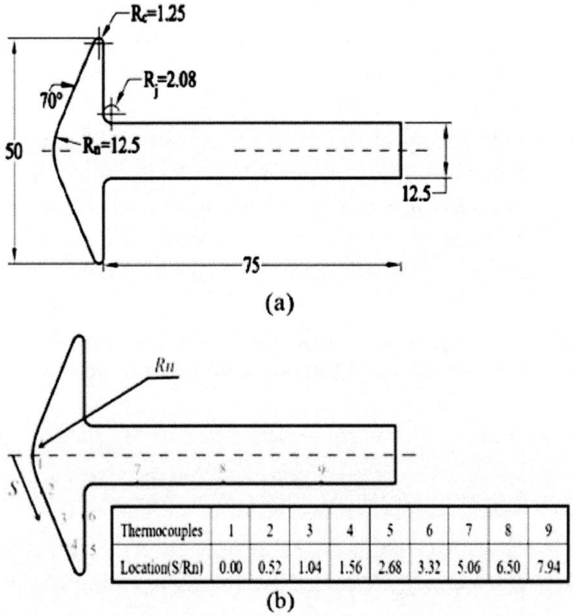

Fig. 1. Configuration of the blunt cone vehicle and thermocouple locations along the test model

Figure 2 show the heat flux at the windward of the blunt cone is high, and the peak heat at the stagnation point is about 10–20 kw/m^2. At the shoulder area of the model, the heat flux drops sharply and the SPARTA's results is lower than the experiment. But the overall consistency of the data is good, which proves the correctness and reliability of the SPARTA to a certain extent.

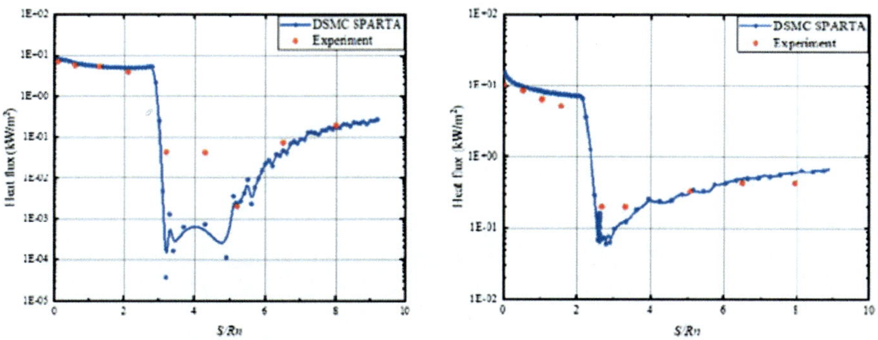

Case A Case B

Fig. 2. Comparison of SPARTA heat flux calculated results and experimental values

3 Simulation Cases and Grid Independence

3.1 Geometry Definition and Freestream Condition

Define the length of the cavity as *L*, the depth as *D*, the length of the plate upstream of the cavity as *Lu*, and the length of the downstream as *Ld*. *D* is fixed at 10 mm, the length-to-depth ratio $L/D = 6$, and the schematic diagram of the cavity model is shown in Fig. 3. The origin of the coordinate system coincides with the left vertex of *Lu*, and the positive directions of the X-axis and Y-axis are set to flow direction and vertical upward respectively. In order to facilitate the analysis of the heat flux, the upstream surface of the cavity is denoted as S1, the bottom surface is denoted as S2, and the downstream surface is denoted as S3.

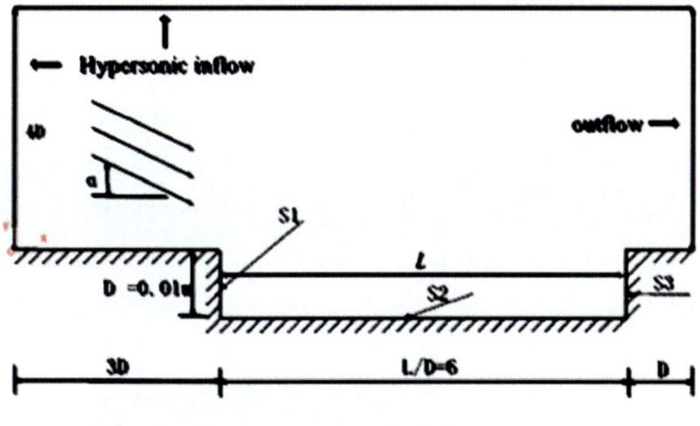

Fig. 3. Geometric sketch of the 2-D cavity

The altitude selected in this paper is 60 km, the free stream Mach number is 8, and other parameters are all determined by American Standard Atmosphere (1976), as shown in Table 2. The angle of attack α is set to seven situations: 0°, 10°, 20°, 30°, 40°, 50° and 60°. Knudsen number $Kn = 0.026$, at this time the model is in the slip region of rarefied flow.

Table 2. Hypersonic free stream conditions at 60 km

Parameter	Value
Altitude (Km)	60
Velocity (m/s)	2520
Density (Kg/m^3)	3.097×10^{-4}
Temperature (K)	247.02
Pressure (Pa)	21.959
Mean free path (m)	2.62×10^{-4}
Number density (m^{-3})	6.4387×10^{21}

3.2 Grid Independence Analysis

The scale of the standard grid is set to 1/3 of the mean free path, so the first-level global grid is divided into 765 × 251, and the second-level grid uses adaptive technology to perform 2 × 2 densification for those that do not meet the requirements of the mean free path. The time step is set to 5×10^{-9}, and the sampling time is 160,000. The number of simulated molecules in each grid keeps at 10 ~ 30. The gas consists 78% N2 and 22% O2. The cavity model with L/D = 6 and $\alpha = 0°$ is taken as an example to test the grid independence with three grid scales: coarse grid (385,192 mesh cells), standard grid (768,060 mesh cells) and refined grid (1,584,048 mesh cells). It can be seen from the Figs. 4 and 5 that three sets of the grid have similar results. so the following will select the standard grid to reduce the computational cost.

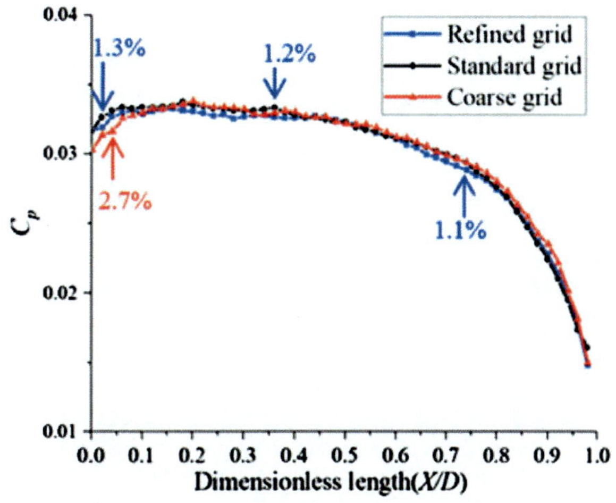

Fig. 4. The pressure distribution of downstream plate of the cavity

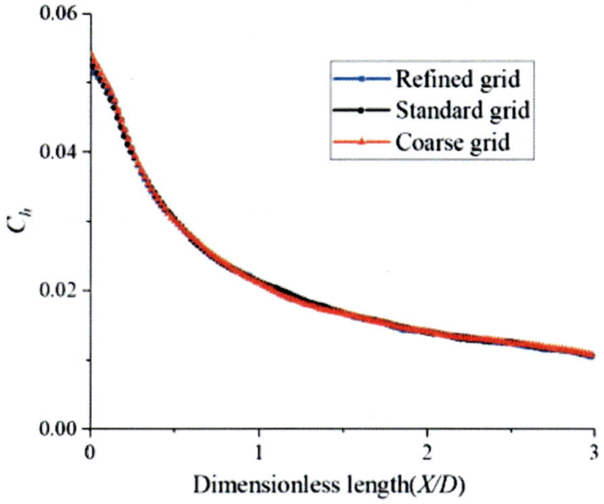

Fig. 5. The heat flux distribution of upstream plate of the cavity

3.3 Simulation Particles Independence Analysis

After setting the standard grid, the area of each grid cell is determined, and the molecular number density of the atmosphere at 60 km is constant, so the number of molecules in the grid cell at the initial time is fixed. Therefore changing the Fnum can increase or decrease the number of simulated particles in the grid cell. The Fnum is setting 1.3×10^{12}, 9.9×10^{11}, 8.2×10^{11}, represents less, standard and more particles respectively. The results of different Fnum conditions have few difference in the field of engineering, a detailed pressure distribution and heat flux distribution of are showed in Figs. 6 and 7.

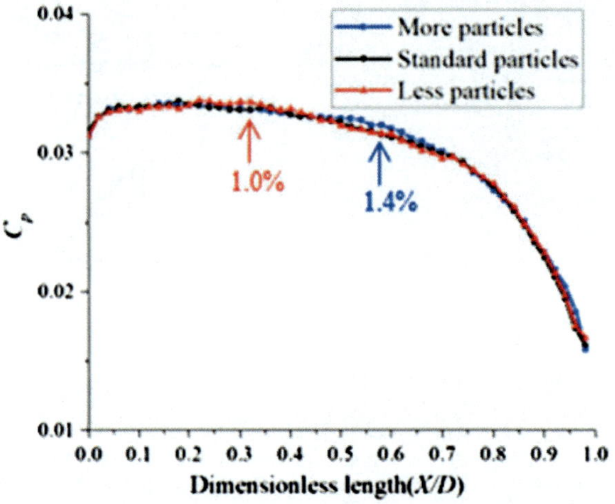

Fig. 6. The pressure distribution of downstream plate of the cavity

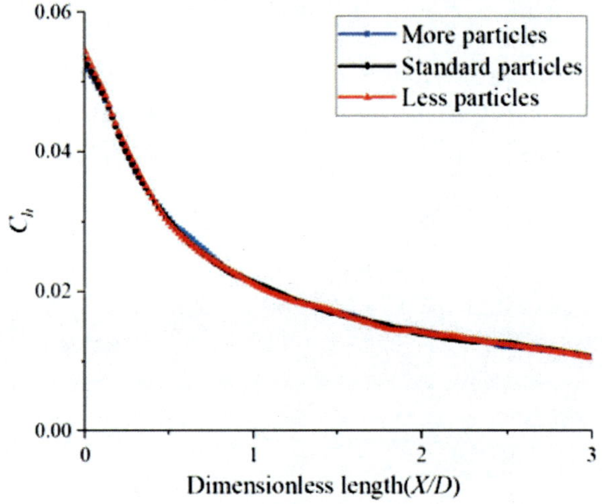

Fig. 7. The heat flux distribution of upstream plate of the cavity

4 Computational Results and Discussion

4.1 Flow Characteristics of the Cavity with Different AOA

Figure 8 shows the Mach cloud diagram of the flow field and a schematic diagram of the recirculation area inside the cavity with $L/D = 6$ changing with AOA α at a height of 60 km and a Mach number of 8. When α increases from 0° to 10°, the free-stream flow cannot enter the bottom of the cavity, the cavity's type changes from closed to open, and the two vortex structures in the cavity gradually become a main vortex structure. When α continues to increase to 40°, an oblique shock wave appears on the plate at the front edge of the cavity, and the free-stream flow with an AOA is deflected into the direction of paralleling the plate after the oblique shock wave. The deflection angle of the shock wave increases with the rising AOA of incoming flow, and the velocity of the flow field after the oblique shock wave decreases. There is still a vortex structure whose core close to the back wall occupies the entire cavity. When the AOA of the incoming flow increases to 50°, a detached shock wave appears on the front edge plate, and the free-stream flow enters the cavity. The original vortex structure has a tendency of becoming a large vortex on the left and a small vortex on the right. The large vortex occupies the length of $4D$ inside the cavity. When $\alpha = 60°$, the incoming flow separates from the front wall, and then attaches to the bottom $2D$ away from the front wall, forming a vortex near the front wall. The flow separates again at $1D$ away from the rear wall, and then attaches to the top of the rear wall, a small corner vortex region is formed near the bottom of the rear wall.

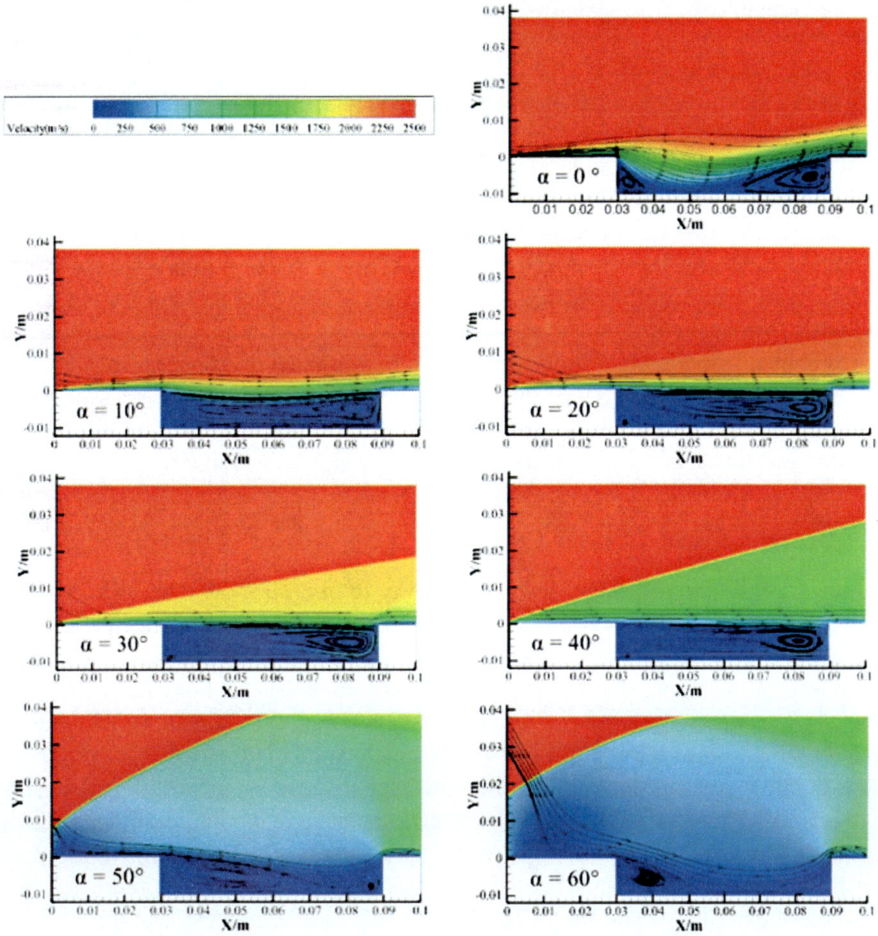

Fig. 8. The Mach cloud diagram of the flow field and a schematic diagram of the recirculation area inside the cavity with different angles of attack

It can be seen from the Mach cloud diagram in Fig. 8 that when the free-stream has an angle of attack, inside the cavity there is a low-velocity region, which is only one-tenth of the speed of the incoming flow. And when AOA increases, oblique shock wave or detached shock wave appears, and the low-velocity region extends to the upper area of the cavity.

Figure 9 shows the pressure cloud diagram of the cavity's flow field. When $\alpha = 0°$, there is a high pressure area after the oblique shock wave in the leading edge of the front plate and another high pressure area near the top of the rear surface because of the reattachment of the shear layer. Because the influence of the oblique shock wave cannot affect so far away and the length of the cavity is too long, the pressure inside the cavity divides two parts: near the front surface is similar to the incoming flow, and near the rear surface is approximately 4 times higher. When $\alpha = 10°$, the high pressure area near the rear surface gradually extends to the front surface, so the pressure inside the cavity increases. And there is 2 times pressure increase behind the oblique shock wave. Due to the expansion of the high pressure area near the rear wall, the distance between the two high pressure areas decreases. When AOA increases, the pressure of near the rear surface still increases, the high pressure area inside the cavity is evenly distributed, and these two parts connect as one region. The max pressure of $\alpha = 20°,\ 30°,$ and $40°$ is about 5.4 times, 9.6 times, and 10.8 times higher than $\alpha = 0°$ respectively. When the AOA increases to 50° and 60°, the oblique shock wave of the leading edge becomes detached shock wave, the value of the high pressure region is still rising, but the pressure of the flow field near the rear surface decreases. So the maximum pressure of the flow field is not the area near the rear surface anymore, the area after the shock wave now has the highest pressure.

Combined with the Mach cloud diagram in Fig. 8, When AOA is less than 50°, there is a bow shock wave formed near the reattachment area and an expansion wave formed at the junction of the rear surface and plate. When the AOA increases to 50° or 60°, the reattachment shock wave disappears, and the expansion wave remains.

4.2 Analysis of Heat Transfer Coefficients

The result of the three surfaces inside the cavity in Fig. 10 show that: at front surface S1, the heat flux is higher near the top, and increases with AOA, but it decreases rapidly $0.1D$ away from the top, the rate of heat flux decrease has slowed down, the value of heat flux of $\alpha = 50°$ is higher than $\alpha = 60°$. The maximum of the heat flux of $\alpha = 10° \sim 50°$ increases 13.2 times, 50.8 times, 78.7 times, 102.9 times, 146.3 times, 190.5 times than no AOA situation. At the bottom surface S2, the heat flux is divided in two situations: one is the flow crossing the cavity which corresponds to $\alpha = 10° \sim 40°$, the heat flux increases with AOA and maximum value appears near the rear point; the other is the flow entering the cavity, corresponding $\alpha = 0°,\ 50°,$ and $60°$, the maximum value appears near the middle of surface and $\alpha = 50°$ is higher than $\alpha = 60°$. At rear surface S3, the trend of heat flux is similar to S1, but $\alpha = 40°$ has the highest heat flux, $\alpha = 50°$ is between the value of $\alpha = 20°$and$\alpha = 30°, \alpha = 60°$ is between the value of $\alpha = 10°$ and $\alpha = 20°$. Among the three surfaces, the top of the rear surface has the highest heat flux. And the heat flux of three surfaces are all sensitive to the change of AOA, so it should be pay more attention if the aerospace vehicle is in such situation.

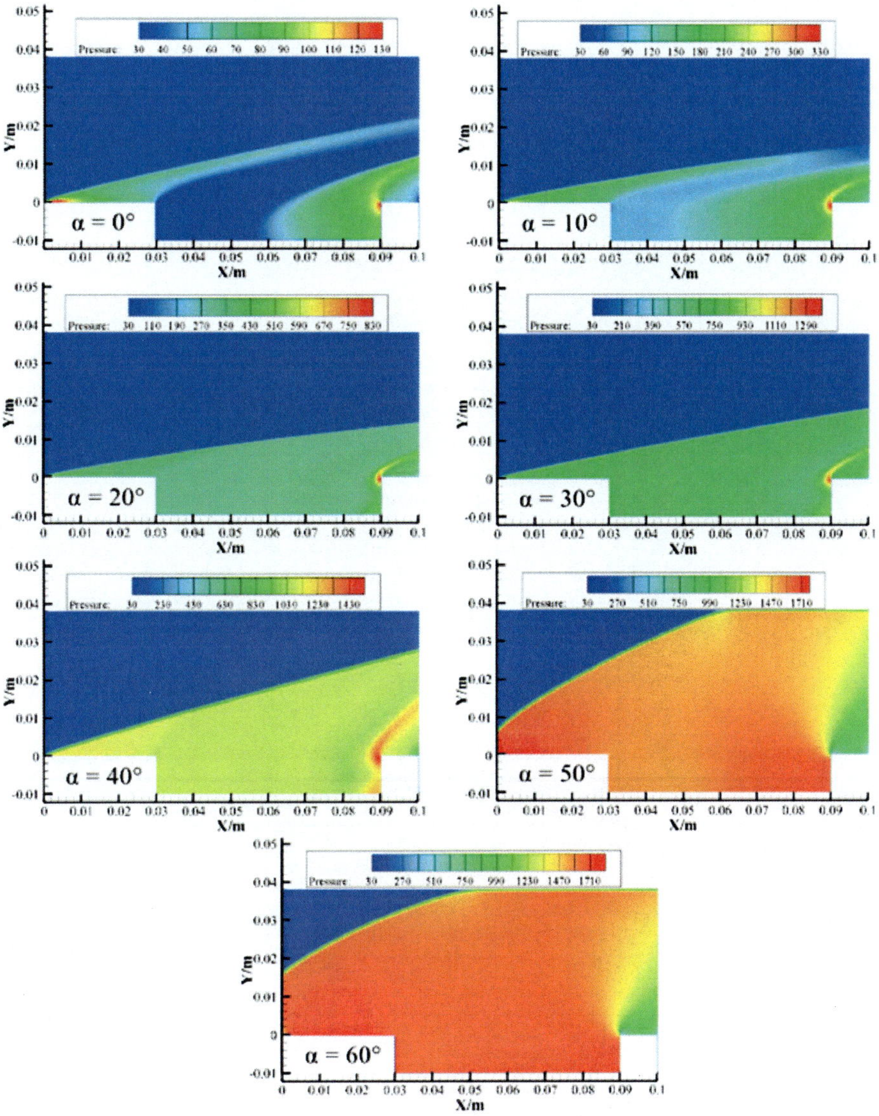

Fig. 9. The pressure cloud diagram of the cavity with different angles of attack

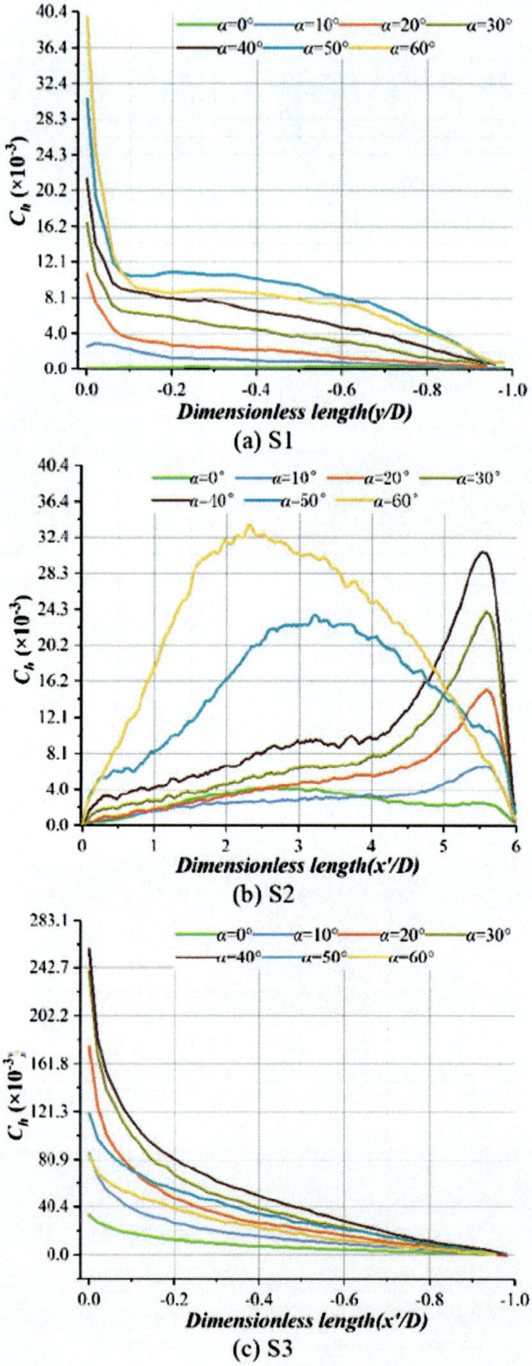

Fig. 10. The heat transfer coefficient distribution with three surfaces of the cavity

5 Conclusions

The direct simulation Monte Carlo (DSMC) method was employed to study the effects of angles of attack on closed cavity, focus on the flow characteristics and the analysis of heat flux. In this paper, AOA ranging from $\alpha = 0°$ to $\alpha = 60°$ and there are some significant conclusion can be obtained, as follows:

(1) The type of cavity changes when AOA changes. When $\alpha < 50°$, free-stream crosses the cavity and there is a vortex occupying the entire cavity. When $\alpha > 50°$, flow enters the cavity and the cavity is back to the closed type when $\alpha = 60°$.
(2) When the incoming flow has an angle of attack, the hypersonic free-stream strikes the plate on the front edge of the cavity and forms a shock wave. When $\alpha < 50°$, there is an oblique shock wave that prevents the stream from entering the cavity. When $\alpha > 50°$, there is a detached shock wave and the flow can enter the cavity.
(3) The low-speed and high-pressure region inside the cavity extends to the upper area of the cavity when AOA increases. Because the flow crosses the shock wave, its speed will decrease and its pressure will increase.
(4) The three surfaces of the cavity are sensitive to the change of AOA. When $\alpha < 40°$, the heat flux increase with rising AOA, but it will decrease when $\alpha > 50°$. The rear surface has the highest heat flux among the three surfaces.

References

1. Petley DH, Smith DM, Edwards CL et al (1984) Surface step induced gap heating in the shuttle thermal protection system. J Spacecraft Rockets 21(2):156–161
2. Gongyue T, Guiqing J (1996) Gap flow analysis and engineering calculation of thermal environment. Chinese J Space Sci Technol V16(6):1–7
3. Bo Q, Yijun G, Haoyuan Z, Lei Z, You'an S, Yewei G (2016) Influence of inflow parameters on vortex structure and thermal environment of heat shield. Acta Aeronaut Sinica 37(03):761–770
4. Charwat AF, Roos JN, Dewey FC, Hitz JA (Jun. 1961) An investigation of separated flows - Part I: the pressure field. J Aerosp Sci 28(6):457–470
5. Charwat AF, Dewey CF, Roos JN, Hitz JA (Jul. 1961) An investigation of separated flows-Part II : flow in the cavity and heat transfer. J Aerosp Sci 28(7):513–527
6. Irving W, Avery Don E, Andrew C (1975) Aerodynamic heating to the gaps and surfaces of simulated reusable surface insulation tile arrays in turbulent flow at Mach 6.6. J. NASA TM-X-3225
7. Avery DE (1978) Aerodynamic heating in gaps of thermal protection system tile arrays in laminar and turbulent boundary layers, NASA TP 1187
8. Binggang T (1990) Discussion on aerodynamic heating of space shuttle thermal shield gap. Aerodyn Exp Meas Control 04:1–8
9. Chun S, Xinlin X, Zhanwei C, Mingxing Y (2012) Analysis of overall flow and heat transfer characteristics of slot cavity seal structure under high speed airflow impact. Acta Aeronaut Sinica 33(01):34–43
10. Bo Q, Haoyuan Z, Yijun G, Lei Z, You'an S, Yewei G (2015) Numerical simulation of vortex structure and aerodynamic thermal environment of hypersonic vehicle. Acta Aeronaut Sinica 36(11):3515–3521

11. Hongming G, Jingqiu C, Li L et al (2015). Experimental study on thermal environment characteristics of thermal barrier tile gap under turbulent conditions. Exp Fluid Dyn 000(002):13–18, 25
12. Guiming T (2000) Experimental study on heat flux distribution in narrow gap. Hydrodyn Exp Meas 04:1–6
13. Santos W , Leite P (2013) Computational analysis of rarefied hypersonic flow over a forward-facing step. Aiaa J
14. Palharini RC, Santos WFN (2019) The impact of the length-to-depth ratio on aerodynamic surface quantities of a rarefied hypersonic cavity flow. Aerospace Ence Technol 88(MAY):110–125
15. Palharini RC, Scanlon TJ, Reese JM (Sep. 2014) Aerothermodynamic comparison of two- and three-dimensional rarefied hypersonic cavity flows. J Spacecraft Rockets 51(5):1619–1630
16. Leite PHM , Santos WFN (2019) Computational analysis of a rarefied hypersonic flow over backward-facing steps. J Thermophys Heat Transfer 1–13
17. Guo G, Luo Q (2018) DSMC investigation on flow characteristics of rarefied hypersonic flow over a cavity with different geometric shapes. Int J Mech Sci 148:496–509
18. Xuhong J, Fei H, Xiaoli C et al (2019) Molecular simulation of gap flow structure and aerodynamic thermal environment of hypersonic vehicle in rarefied flow region. Acta Aerodyn Sinica 34(001):201–209
19. Junyan Z, Xuede W (2020) DSMC study on the geometric characteristics of near space hypersonic cavity. Acta Ballistics 32(01):47–54
20. Bird, GA (2015) Aerodynamic properties of some simple bodies in the hypersonic transition regime. Aiaa J 4(1):55–60
21. Allegre J et al (1997) Experimental rarefied aerodynamic forces at hypersonic conditions over 70-degree blunted cone. J Spacecraft Rockets 34(6):719–723
22. Roohi E, Stefanov S (2016) Collision partner selection schemes in DSMC: from micro/nano flows to hypersonic flows. Phys Rep 656:1–38

New Engine Design and Development

The Cause and Effect Analysis of Aero-Engine Roller Bearing Skid Based on Assembly Process Inspection and Trial Run

QingZhen Gao(✉) and DaoYuan Pan

School of Mechanical Engineering, Anhui Polytechnic University, Wu Hu 241000, P.R. China
gaoqingzhen68@sohu.com, plongroad@163.com

Abstract. In this paper, firstly the reasons of aero-engine bearing skid were analyzed from these aspects of fault phenomenon, skid mechanism, manufacturing process, practical use, trial run, measurement and inspection by fishbone diagram. In order to ensure the economy and airworthiness of aero-engine manufacturing, the use and repair of bearing with skid damage were discussed by means of information statistics and trial run experiments. Then the effect of bearing skid on the engine safety was analyzed. Based on the above research, the possible causes and preventive measures for bearing skid are expounded. And the repair method, experimental content and inspection standard of bearing with skid are presented in detail. The research has important guiding significance for aero-engine bearing to master its working state, prevent its fault and guarantee the aviation safety.

Keyword: Roller bearing skid · Manufacturing process analysis · The effect on aero-engine · Trial run

1 The Mechanism of Bearing Skid

As shown in Fig. 1, The bearing skid is an inherent problem for an engine and directly related to the engine structural design [1].

The roller bearing is mounted on the turboshaft. When a bearing works, all rollers should roll purely on the inner and outer ring raceways [2]. To make the roller roll purely, there must be sufficient dragging force between the raceway and every roller to overcome the resistance that hinders the normal movement of the roller-retainer combination. Otherwise, the bearing cannot move according to the relationship between the planetary gear system [3].

The sliding friction will be formed between every roller and the raceway. As the sliding friction coefficient is greater than the rolling friction coefficient and some external factors make some rollers form relative friction between the inner and outer ring raceways that cause skid and scratch marks and local surface wear on the inner and outer ring raceways, which is called "skid mark". Theoretically the relative movement between all rolling bearing components should be pure rolling at all contact spots.

That is so-called that "the linear velocities of two pure rolling components are equal at the contact spot" [4]. Otherwise, it is not pure rolling but skid. The skid does not

H. Ding (ed.), *Aerospace Mechatronics and Control Technology*, Springer Aerospace Technology,
https://doi.org/10.1007/978-981-16-6640-7_5

Fig. 1. The bearing skid mark

necessarily result in failure. However, once a bearing skids, that is, there is a relative sliding at the contact spot, it may cause sliding damage on the contact surface, which is called "slide or skid". But, there will not definitely be sliding failure when skid. Skid just sets the stage for the sliding failure. On the contrary, a bearing with sliding damage must have skidded.

2 The Cause Analysis of Skid

2.1 The Engine Manufacturing Process Analysis

As shown in Figs. 2 and 3, the fishbone charts were drawed to find the possible factors of bearing skid from five aspects of human, machine, material, method and environment. Then these factors are analyzed one by one in Table 1.

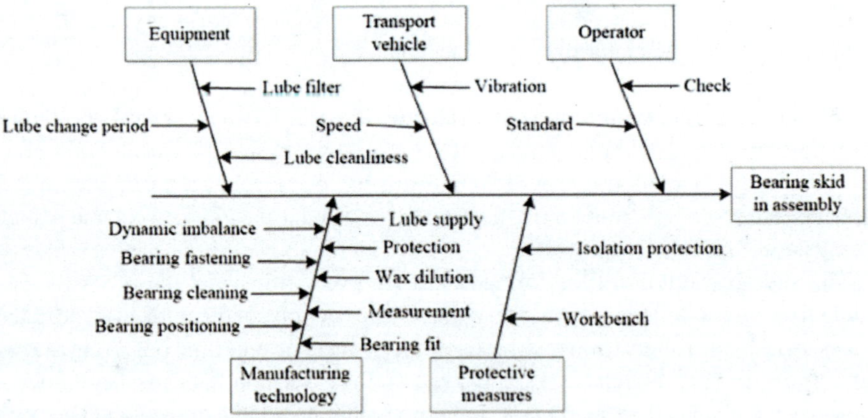

Fig. 2. Fishbone diagram for assembly

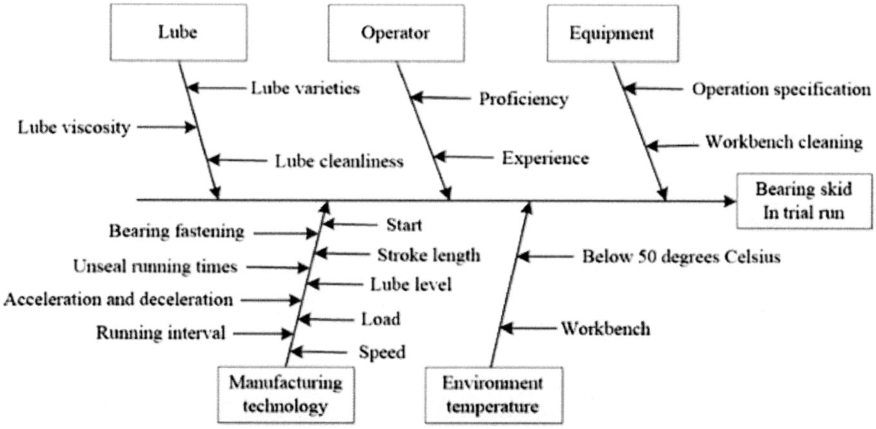

Fig. 3. Fishbone diagram for trial run

2.2 The Investigation of Bearing Skid

By consulting some institutes and factories, the similar skid phenomena also occurred in the bearings. But it does not affect the bearing use.

In order to reduce bearing skid, during the assembly process, the radial clearance of bearing, the interference of inner and outer steel sleeves and cleanliness such as beeswax and vaseline shall be controlled strictly. when measuring the roundness of outer shaft bushing and the inner diameter of inner steel sleeve and, the positions of the long and short shafts must be marked; When installing the inner steel sleeve, its long shaft shall be aligned with the outer diameter of outer shaft sleeve. When the steel sleeve is cooled, the roundness of inner steel sleeve shall be not more than 0.005 mm. When measuring the roundness of inner hole on bearing seat, the position of long axis and short axis should be marked. When installing the outer steel sleeve, its long shaft shall be aligned with the inner diameter of bearing seat. The combined clearance of bearing after installation is controlled from 0.10 to 0.15 mm. And the number of measurement during bearing selection, roundness measurement method, nut tightening etc. should meet the requirements.

The inspection standards for bearing skid from repair factory are as follows: (a) Single scratches are allowed on the raceway, not exceeding 0.15 mm in width, not exceeding 0.03 mm in depth, and not exceeding a quarter in length. The axial scratches on raceway are not allowed. After deburring on bearing with ointment soaked felt, the bearing can continue to be used. (b) The number of rust spots on raceway working surface shall not exceed 3, each rust spot shall not exceed 0.2 mm in diameter and 0.05 mm in depth. The number of rust spots on raceway non-working surface shall not exceed 5, each rust spot shall not exceed 0.3 mm in diameter and 0.1 mm in depth. (c) The number of discontinuous distributed pits on inner and outer steel sleeve raceway shall not exceed 3, each pit shall not exceed 0.3 mm in diameter and 0.005 mm in depth. (d) The bearing roller is allowed to have a single scratch with width less than 0.2 mm and depth less than 0.1 mm in the circumferential direction. Or two scratches with spacing greater than 3 mm, each with width less than 0.1 mm and depth less than 0.05 mm.

Table 1. The failure cause analysis for bearing skid

Number	Symbol	Fault tree analysis [5]	Project analysis	Proposed measures
1	DH11	Manufacturing process	Work reliability	Reinspection and Information sharing
2	DH31	Bar difference [6]	Work reliability and Cleanliness control	Pre-assembly inspection
3	DH32	Equipment	Work reliability	Maintenance
4	DH33	Radial clearance	Work reliability and Cleanliness control	Adding measurement points
5	DH34	Mating the inner steel sleeve and shaft	Work reliability	Controlling the dimensions of relevant parts
6	DH35	Roundness of bearing steel sleeve after assembly	Work reliability	Controlling the dimensions of relevant parts
7	DH36	Rotor unbalance [7]	Dragging force of bearing rollers	Meeting the equilibrium requirement
8	DH41	Equipment	Cleanliness control of lubricating oil system	Controlling oil grade and Maintenance
9	DH42	Assembling tightness	The stress on bearing	Controlling the fastening value and Reinspection
10	DH43	Concentricity [8]	The stress on bearing	Controlling assembly process
11	DH44	Vaseline	The application amount	Diluting with hot oil and Increase rotor rotation
12	DH45	Nozzle flow	Lubrication and cleanliness	Increase oil flow speed
13	DH46	The operators	Operation proficiency and experience accumulation	Increase training
14	DH51	Oil	Viscosity and cleanliness	Controlling oil grade
15	DH52	Test environment	Cleanliness	Maintaining cleanliness
16	DH54	The operators	Operation proficiency and Experience accumulation	Increase training

(continued)

Table 1. (*continued*)

Number	Symbol	Fault tree analysis [5]	Project analysis	Proposed measures
17	DH53-1	Starting trial run	Starting time and bearing lubrication effect	Manually rotating rotor before starting
18	DH53-2	The process of trial run	Bearing lubrication effect	Avoiding running the engine at a low speed for a long time
19	DH53-3	Lubricating oil pressure	Lubricating oil supply	Improving oil system efficiency
20	DH53-4	Engine unsealing and running	Bearing lubrication effect	Manually rotating rotor before starting
21	DH54-5	Acceleration and deceleration of engine	Bearing force	Reinspection of bearing design quality
22	DH53-6	Rotational speed and load of engine	Bearing force	Reinspection of bearing design quality

The inspection results of 5 skid bearings (Serial number:9-1-66, 9-1-31, 9-1-61, 9-1-51, 9-1-53) are shown in Tables 2 and 3. Through comparison, it is found that the roughness of skid area is higher than that of non-skid area, and the hardness has no change.

The datas from 100 engines with bearing skid are analyzed. The datas include bearing assembly parameter, staff ability, season, bearing batch, engine type etc. The calculation results show that the statistical value P is greater than 0.05, there is no significant difference between the above parameters, and the key factor of bearing skid is not found.

The engine trial runs show that the bearing skid occurs during the engine starting phase. It is preliminarily believed that bearing skid may be due to a higher spot on steel sleeve raceway. That is to say, The roundness value on working face of bearing inner steel sleeve is relatively larger, then skid occurs during working. If the higher spot is removed, the occurrence possibility of bearing skid will be reduced. The above inference has some rationality through some experiments.

3 the Effect of Bearing Skid

3.1 The Engine Service Status Statistics

In order to further study whether the bearing after polishing repair has an impact on engine reliability. The usage data of 96 engines with repaired bearing were collected. Among them, 37 engines have been scrapped, 13 engines have a service life of more than 200 h, 24 engines have a service life of between 100 and 200 h, and 22 engines have a service life of less than 100 h.

According to the statistics, these engines are in good use and there is no failure due to bearing skid. So the method of polished repair on skid bearing is feasible, and the engine reliability will not be affected by polished bearing.

Table 2. The inspection records of skid bearings

Project		Standard values	Unit	9-1-66	9-1-31	9-1-61	9-1-51	9-1-53
△dmp		0 ~ −13	μm	−6	−10	−7	−6	−11
Vdp		13		3	4	3	3	2
Vdmp		7		1	2	1	3	0
△Dmp		0 ~ −13		−5	−5	−12	−7	−7
VDp		13		2	2	3	3	3
VDmp		7		2	1	1	1	0
Gr		200 ~ 230		214	215	216	220	213
Roller	Group difference	1		0.6	1	0.5	1	1
	Ellipse	0.4		0.25	0.25	0.2	0.25	0.2
	Height difference	7		5	8	5	5	5
Outer ring raceway size		80 ~ +120	mm	+75	+100	+60	+110	+100
Outside diameter of retainer		166.05 ~ 166.20		166.18	166.20	166.19	166.17	166.19
Guiding clearance				0.893	0.9	0.87	0.945	0.91

Table 3. The measurement records of bearing roughness and hardness

Project	9-1-66	9-1-31	9-1-61	9-1-51	9-1-53	Standard value
Ra (Sliding zone)	0.133	0.09	0.11	0.094	0.152	0.125
Ra (Non-sliding zone)	0.049	0.046	0.058	0.055	0.057	
Inner raceway (HRc)						
Inner ring hardness of no. 9-1-61 bearing	Instrument	Sliding zone	The transition zone between sliding zone and non-sliding zone		Non-sliding zone	Terminal face of inner raceway (HRc)
	Vickers hardness	61.5	61.5		61.5	61.2
	Rockwell hardness	62	62.2		62	62

3.2 The Trial Run of Polished Bearing

For a bearing, the free clearance is 0.23 mm, the fit interference between inner steel sleeve and outer shaft bushing is 0.03 mm and between inner steel sleeve and bearing seat is 0.07 mm. The combined bearing clearance is 0.14–0.15 mm. After the trial run, there are two skid marks on the surface of bearing inner steel sleeve, one mark is 80 mm long and 2 mm wide, and the other mark is 60 mm long and 1 mm wide. The skid surface was trimmed. After an additional trial run, the bearing had no skid fault.

And it was disassembled and inspected after a long-term trial run with a total working time of 370 h 50 min. As shown in Fig. 4, there is a scratch with a width of about 1.0–1.5 mm and a length of about 85 mm on the outer diameter of inner steel sleeve.

Fig. 4. Bearing skid after long-term trial run

From the above tests, the engine works normally and all parameters are stable. No abnormal situation occurred due to bearing skid. However, it is not clear when the skid will occur and whether it will affect the engine life.

3.3 The Bearing Inspection After Trial Run

After trial run, the bearing is decomposed and checked. The result shows that the roundness and roughness of inner steel sleeve and outer steel sleeve are less than 0.005 and 0.125 mm. The bearing combination clearance is 0.15 mm. These values meet the specified requirements as shown in Table 4. The inner and outer steel sleeve are brown yellow and the wear marks on them are normal. The color of bearing retainer and rollers is dark brown, they are worn slightly and evenly.

Table 4. The roundness and roughness values

Project	Test spot 1	Test spot 2	Specified value
Roundness (μm)	2.33	2.262	5
Roughness (μm)	0.1071	0.1130	0.125

4 Conclusion

From the above analysis, the key cause of bearing skid has been found. And the probability of bearing skid can be reduced by controlling the radial clearance, the interference degree of inner and outer steel sleeve and the cleanliness in assembly process. The aero-engine 300 h trial run with polished bearing shows that bearing skid is not regular. It is not certain whether the skid will reappear during the trial run. But the polished bearing will not affect the safe and reliable work for an engine within 300 h.

5 Suggestions and Measures

The polishing bearing can be used according to the test results. Considering the uncertainty of bearing skid, It is suggested that the bearing with skid mark should be treated as follows.

(1) If the skid area does not exceed 1/2 of bearing raceway surface and no roller is damaged, no additional test is required for polished bearing. If the skid area exceeds 1/2 but does not exceed 2/3 of bearing raceway surface, and no roller is damaged, the skid bearing needs to be tested again after polishing repair.
(2) The specific repair methods of bearing skid are as follows. (a) Marking the bearing skid range firstly. (b) Removing scratches with oil-soaked sandpaper. (c) Polishing the scratch area with oil-soaked metallographic sandpaper. The polishing range is slightly larger than the marking range. (d) Cleaning the bearing with gasoline after polishing.

Acknowledgements. This work is supported by the 2019 Talent Fund Project of Anhui Polytechnic University (No. 2019YQQ001) and the cultivation project of Anhui Polytechnic University (Xjky2020005).

References

1. Xiangyu X, MingWei F (2017) Analysis of sliding and rubbing damage morphology of aero-engine bearing. J Bear 32(9):52–54 (Chinese)
2. Li J, Chen W, Xie Y (2014) Experimental study on skid damage of cylindrical roller bearing considering thermal effect. J Proc Inst Mech Eng Part J J Eng Tribol 228(10):1036–1046
3. Wang Y, Wang W, Zhang S (2015) Study of skidding presents in contact ball bearings under high speed. J Tribol Int Edn 92(4):404–417
4. Bin W, ZhaoYuan F (2017) Analysis of sliding friction fault of aero-engine spindle bearing. J Harbin Bear 38(1):15–18 (Chinese).
5. ZiBin W, SiEr D (2019) Analysis of running stability of high-speed cylindrical roller bearing retainer. J Vibr Impact 38(9):48–50 (Chinese)
6. FanHong (2019) Experimental and simulation analysis of full roller bearing torque. J Meas Test Tech 46(4):26–30 (Chinese)
7. YaQiong Z, Yan S (2014) Variation rules of the corrugated degree and roughness morphology of cylinder liner wear surface. J China Surf Eng 27(5):23–25(Chinese)
8. Di Z (2019) Tribological analysis of surface texture in rolling bearing system. J Bear 58(3):30–34 (Chinese)

Dynamic Detection Method for Blade Tip Clearance of Aero-Engine in Assembled State Based on Linear Structured Light

Gang Zhao[1], Maocheng Xu[1], Wenlei Xiao[1], Hui Wang[2], Bing Zhang[3], and Pengfei Zhang[1](✉)

[1] Beihang University, Beijing, China
ftd423@buaa.edu.cn
[2] Tsinghua University, Beijing, China
[3] Aero Engine Corporation of China, AECC Shenyang Engine Research Institute, Beijing, China

Abstract. Aero-engine blade tip clearance (BTC) is one of the most important parameters that have decisive influence on engine performance. Thus, it is a typical inspection indicator during engine assembly. With the higher demand for consistency of engine performance, the measurability of BTC is required to be much more accurate and dynamically. However, the traditional BTC detection method in the assembly process is restricted to lower accuracy, inability of complex structures and lack of dynamic measuring capability. This paper introduces a visual BTC measuring system based on structured light, which can directly measure the clearance between rotor and stator casing by observing its position from the engine principle axial direction. A special laser generator is designed to project structured light to the surfaces of blade and casing. Simultaneously, the light is captured by the industrial camera. Compared to the traditional methods, the BTC in whole circumferential phase can be measured rather than single phase. Meanwhile, the proposed method is not influenced by casing deformation and installation errors. Finally, the feasibility and capability of the system is verified by simulation experiment.

Keywords: Aero-engine · Tip clearance · Dynamic measurement · Line structured light · Visual measurement

1 Introduction

The blade tip clearance (BTC) is an important index that affects the performance of aero-engines. The value of BTC directly affects the vibration and aerodynamic performance of the engine. Thus, it is an important technologic parameter in the engine assembly process, which reflects the final assembly result of the rotor and the casing. The measurement result of BTC is an important value used to control and evaluate the assembly quality of the engine.

At present, the measurement method of aero-engine blade tip clearance has been widely studied, which can be mainly divided into opening hole measurement method,

H. Ding (ed.), *Aerospace Mechatronics and Control Technology*, Springer Aerospace Technology,
https://doi.org/10.1007/978-981-16-6640-7_6

half-casing measurement method, virtual assembly evaluation method, and blade tip sensor method. Among them, the opening hole measurement method is widely studied. This method is measured by opening a hole in casing which is designed to fix the sensor directly above rotor blade. It has advantage of good sensor compatibility and good real-time. The sensors used mainly include discharge probes [1], capacitance [2], inductance [3], eddy current [4], optical fiber [5, 6] and microwave sensing [7]. It can realize the measurement of the rotor at high speed [8] and high temperature [9]. Further, researchers develop tip timing [10] and event capture [11] to make this method more reliable. But it can only obtain the changing rule of the BTC at one fixed circumferential phase in casing. However, in order to ensure the quality of the product, large-scale BTC measurement holes will not be allowed in the factory production process, so this method is not suitable for real production conditions. The half-open casing measurement method [12] means that BTC between the rotor and the half stator casing can be measured at the position of casing mounting edge in assembly process, when only a half stator casing is installed. At the same time, because of the higher rigidity of the casing mounting edge, it is difficult to reflect the BTC in other positions. The virtual assembly evaluation method records the casing run-out and rotor run-out parameters during the assembly process, and uses these as data basis to predict the BTC after the assembly being completed. This method takes less consideration of the supporting structure and bearing. But the application is relatively simple and feasible. Supposing that the support structure and the bearing meet the requirements, we can use assembly analysis methods such as Monte Carlo method to predict BTC value after assembly. The blade tip sensor method mainly includes Linipot [13] and Linigage systems. They fix non-contact sensor on a blade and wind the lead wire around the rotor. After assembly, workers rotate the rotor and record sensor readings. At the same time, the lead wire and sensor is taken out, so the BTC can only be measured once. This method cannot directly obtain the tip clearance value because the relative position of the sensor to the blade is unknown. This method intuitively reflects the run-out of the tip clearance. Due to the introduction of a new tooling, so how too taken it out is very worth considering.

For the BTC measurement in engine assembly process, this paper proposes a visual measurement method of BTC based on line structured light. The non-contact measurement method is used to measure BTC from the engine principle axial direction. It can realised be fixed on any circumferential phases of the stator casing to measure BTC value. Finally, the feasibility and capability of the system is verified by simulation experiment.

2 The Design of Blade Tip Clearance Detection System Based on Line Structured Light

2.1 Overview of the Measurement System

For accomplishing measurement of BTC, the scalability and stability of the measurement need to be satisfied, and the measurement system is required to be able to measure multiple phases and scenarios. It should not be affected by deformation of the casing and installation error, so structured light and industrial camera is used to observe in axial direction. The principle of BTC measurement is based on line structured, and the system mainly includes line structured light emitter, industrial camera, fixed tooling and so on (Fig. 1).

Fig. 1. System components.

2.2 Line Structured Light Measurement Principle

In line laser measurement, the laser emits a laser plane to illuminate the surface of the object to be measured. The reflected light is converged at another angle through the imaging system, and the linear light bar will be imaged on the camera sensor. When the surface of the measured object changes, the image formed by the light bar on the camera would also change accordingly. The light bar in image and actual world has a unique correspondence relationship. By calculating the corresponding relationship, we can figure out where the light bar is in the real world. The specific measurement principle is depicted as follows (Fig. 2).

The following four coordinate systems are involved in the figure.$O_w - X_wY_wZ_w$ is world coordinate system, describing the camera position. $O_c - X_cY_cZ_c$ is camera coordinate system, optical center is origin. $o - xy$ is image coordinate system, the origin is the midpoint of the imaging plane. uv is pixel coordinate system, the origin is the upper left corner of the image (in openCV). P is a point in the world coordinate system, namely is a real point in life. p is the imaging point of point P in the image. The coordinates in the image coordinate system are (x, y), and the coordinates in the pixel coordinate system are (u, v). f is camera focal length, equal to the distance between o and O_c, $f = \|o - O_c\|$.

Through coordinate transformation and other methods, we can get the conversion method from the world coordinate system to the camera coordinate system, so the conversion equation of point P is

$$\begin{bmatrix} X_c \\ Y_c \\ Z_c \end{bmatrix} = R \begin{bmatrix} X_w \\ Y_w \\ Z_w \end{bmatrix} + T \tag{1}$$

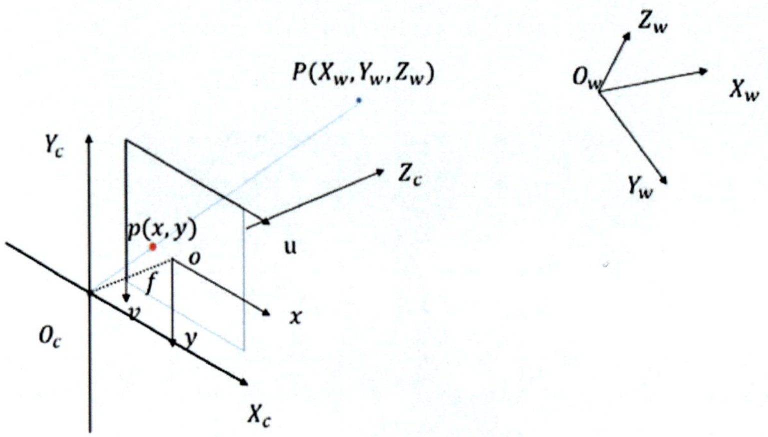

Fig. 2. Camera imaging principle.

That is

$$\begin{bmatrix} X_c \\ Y_c \\ Z_c \\ 1 \end{bmatrix} = \begin{bmatrix} R & T \\ \vec{0} & 1 \end{bmatrix} \begin{bmatrix} X_w \\ Y_w \\ Z_w \\ 1 \end{bmatrix} \tag{2}$$

Among them, $\begin{bmatrix} R & T \\ \vec{0} & 1 \end{bmatrix}$ is noted as the external camera parameter, M_2.

According to the method of similar triangles, from $\Delta\ ABO_c \sim \Delta\ oCO_c$, $\Delta\ PBO_c \sim \Delta\ pCO_c$ we can get

$$\frac{AB}{oC} = \frac{AO_c}{oO_c} = \frac{PB}{pC} = \frac{X_c}{x} = \frac{Z_c}{f} = \frac{Y_c}{y} \tag{3}$$

that is

$$x = f\frac{X_c}{Z_c}, y = f\frac{Y_c}{Z_c} \tag{4}$$

written as a matrix

$$Z_c \begin{bmatrix} x \\ y \\ 1 \end{bmatrix} = \begin{bmatrix} f & 0 & 0 & 0 \\ 0 & f & 0 & 0 \\ 0 & 0 & 1 & 0 \end{bmatrix} \begin{bmatrix} X_c \\ Y_c \\ Z_c \\ 1 \end{bmatrix} \tag{5}$$

Connecting the equation obtained before from the world coordinate system to the camera coordinate system can be obtained

$$Z_c\begin{bmatrix} u \\ v \\ 1 \end{bmatrix} = \begin{bmatrix} \frac{1}{dx} & 0 & u_0 \\ 0 & \frac{1}{dy} & v_0 \\ 0 & 0 & 1 \end{bmatrix}\begin{bmatrix} f & 0 & 0 & 0 \\ 0 & f & 0 & 0 \\ 0 & 0 & 1 & 0 \end{bmatrix}\begin{bmatrix} R & T \\ \vec{0} & 1 \end{bmatrix}\begin{bmatrix} X_w \\ Y_w \\ Z_w \\ 1 \end{bmatrix}$$
$$= \begin{bmatrix} f_x & 0 & u_0 & 0 \\ 0 & f_y & v_0 & 0 \\ 0 & 0 & 1 & 0 \end{bmatrix}\begin{bmatrix} R & T \\ \vec{0} & 1 \end{bmatrix}\begin{bmatrix} X_w \\ Y_w \\ Z_w \\ 1 \end{bmatrix} \tag{6}$$

where $\begin{bmatrix} \frac{1}{dx} & 0 & u_0 \\ 0 & \frac{1}{dy} & v_0 \\ 0 & 0 & 1 \end{bmatrix}\begin{bmatrix} f & 0 & 0 & 0 \\ 0 & f & 0 & 0 \\ 0 & 0 & 1 & 0 \end{bmatrix}$ is recorded as the camera internal reference M_1 so the above formula is expressed as

$$Z_c\begin{bmatrix} u \\ v \\ 1 \end{bmatrix} = M_1M_2\begin{bmatrix} X_w \\ Y_w \\ Z_w \\ 1 \end{bmatrix} \tag{7}$$

Obviously, when a certain point P is known, the u and v of the pixel on the camera can be easily calculated. But when u and v are known, the coordinates of the world coordinate system of point P cannot be calculated, namely it is not possible to calculate the three-dimensional information of an object only through a single image. Therefore, the information provided by binoculars or other auxiliary structured lights is required to register the 2D points to 3D. The line laser is a light plane emitted by laser, which forms a linear light bar in the camera coordinate system when it is irradiated on object. The equation of the line laser plane can be expressed as:

$$AX_w + BY_w + CZ_w + D = 0 \tag{8}$$

After introducing the line laser, the points on the line laser plane which means the points on light bar in the image will satisfy equation:

$$\begin{cases} Z_c\begin{bmatrix} u \\ v \\ 1 \end{bmatrix} = M_1M_2\begin{bmatrix} X_w \\ Y_w \\ Z_w \\ 1 \end{bmatrix} \\ AX_w + BY_w + CZ_w + D = 0 \end{cases} \tag{9}$$

Therefore, after obtaining the parameters (A, B, C, D) of the line laser plane equation and the M1,M2 of the camera, the values of X_w, Y_w, and Z_w can be calculated from the u and v on the image. So the 3D information on light bar can be obtained after system calibration, and the BTC of the engine will be calculated.

2.3 System Calibration

According to the above analysis, for solving the 3D coordinates of the point on the line laser, the internal and external parameters of the camera and the line laser plane equation need to be calibrated. In order to simplify the system calculation process, this paper selects the world coordinate system as the camera coordinate system, so the external parameter matrix of the camera is a unit matrix. By using checkerboard calibration algorithm [14], the overall calibration of the line laser and the camera is completed using similar process (Fig. 3).

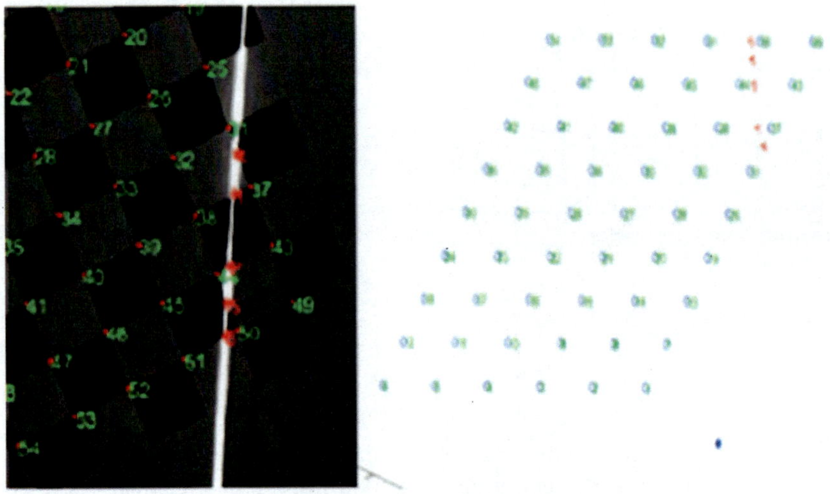

Fig. 3. System calibration method.

According to the constant projection, the proportional line segment in 3D space will maintain the corresponding projection ratio after being mapped to the 2D space. Therefore, we can know the corresponding relationship of the points in the 3D space through the corresponding relationship of the points on the 2D image relationship. Based on the calibration checkerboard, the light pattern formed by the line structured light on the calibration plate will be a straight line. The line between two corners of the calibration plate will also be a straight line. During the calibration process, the corresponding straight line maintains the corresponding proportional relationship, as shown in the figure (Fig. 4).

According to the principle of invariance of projection, $\frac{|\overrightarrow{A_w,B_w}|}{|\overrightarrow{A_w,P_w}|} = \frac{|\overrightarrow{A,B}|}{|\overrightarrow{A,P}|} = \lambda$ is known, then the coordinates of P_w in the space coordinate system can be calculated through the relationship between A, B, P on the image, which can be expressed as:

$$\begin{cases} X_p = \frac{X_A+\lambda X_B}{1+\lambda} \\ Y_p = \frac{Y_A+\lambda Y_B}{1+\lambda} \\ Z_p = \frac{Z_A+\lambda Z_B}{1+\lambda} \end{cases} \quad (10)$$

In this way, the 3D coordinates of the points on the line structured light can be obtained. So the coordinates of all the points on the picture used for calibration can be

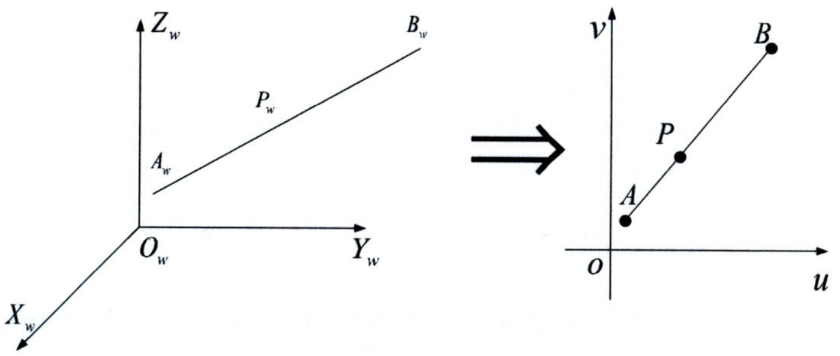

Fig. 4. Line segment mapping

calculated. After a sufficient number of points on the line structured light plane being obtained, the light plane equation can be obtained by plane fitting (Fig. 5).

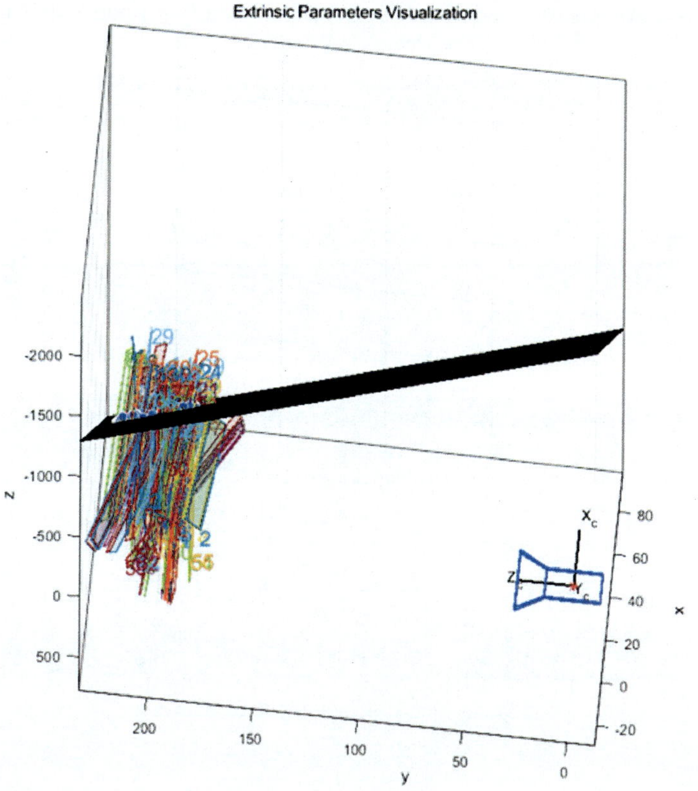

Fig. 5. Calibration result

It is obvious that this calibration method can effectively calibrate the relevant parameters of the measurement system, which provides a basis for the overall measurement system.

2.4 Clearance Recognition and Calculation

For the image obtained by line laser emitter, further processing and calculation are required to obtain the value of the BTC. For BTC recognition and calculation, the main process is performed as follows:

1. Laser centerline extraction.

   (a) The original image is performed with white balance, binarization and filter processing.
   (b) The center of laser stripe is used closed operation to extract skeleton method.
   (c) The laser light stripe area is extracted from the processed image by steger algorithm.
   (d) The line segment formed by the line laser emitter the blade is detected by hough line detection.
   (e) Through identifying and splicing the line segments with similar slopes, the ideal line segments required for 3D restoration is finally formed (Fig. 6).

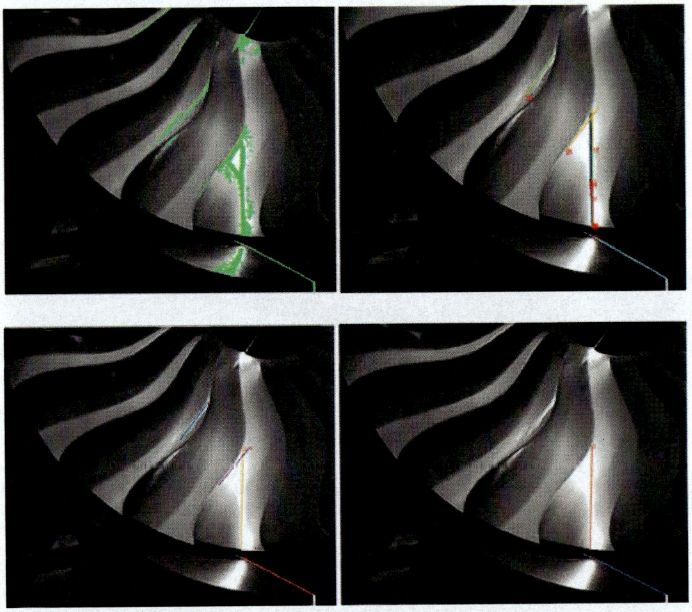

Fig. 6. Line structured light extraction.

2. Definition and calculation of blade tip clearance.

   For each measurement method, it is necessary to define the specific meaning of its measurement standard. Therefore, this article defines the BTC as shown in the figure (Fig. 7).

   In the case of 2D projection, the BTC value of point A on the blade tip is the rotor radius OA extending to point B on the casing. The length of the line segment AB

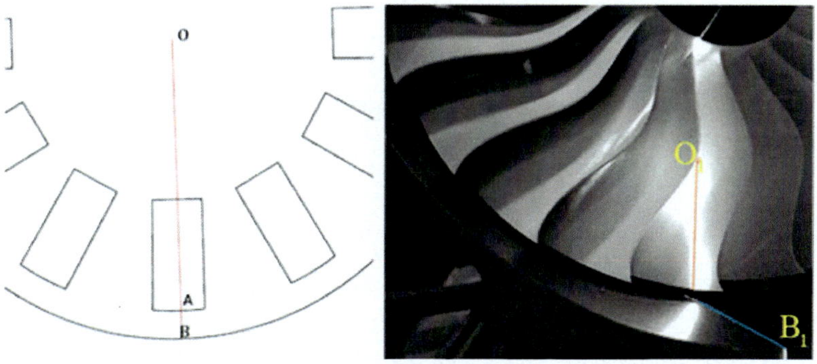

Fig. 7. Schematic diagram of tip clearance.

is the clearance value corresponding to point A. Corresponding to the actual picture on the right, the clearance can be simplified to the distance between the space line segment O1A and BB1 by ensuring that the line laser passes through the engine axis. The BTC value can be calculated by calculating the spatial position relationship of the two space line.

3 Simulation Experiment Verification

Since there is no sufficient accurate means to describe and measure BTC as mentioned above, in order to accurately evaluate the accuracy of the measurement method, the simulation rendering experiment is performed for verification. Through comparing the actual data from the CAD model and measurement data from simulation experiment, we can obtain the preliminary accuracy of the measurement method and verify its feasibility. The 3D model is imported into the KeyShot software, and the line structured light and the camera is set in the appropriate position to build a virtual experimental system (Fig. 8).

In software, the pictures used for calibration and measurement are rendered according to physical priciples. After the system is calibrated, the calculated value of BTC is 4.22 mm, and the clearance value directly measured from the CAD software is 4.2 mm, which means the relative error is 0.47%. It confirms this method satisfies accuracy requirements (Fig. 9).

4 Discussion and Conclusion

In response to the dynamic and full-circumferential measurement of aero-engine BTC. This article introduces a visual BTC measurement method based on line structured light, it includes the principle of the system, corresponding calibration and measurement methods, finally verifies the feasibility of the measurement method in a simulated environment. As a new BTC measurement method, the structured light measurement method needs to be further improved and optimized. It is further combined with the actual engine model to design a circumferential mobile tool, so that it can measure BTC distribution along the entire circumference of stator casing. With detailed picture brought

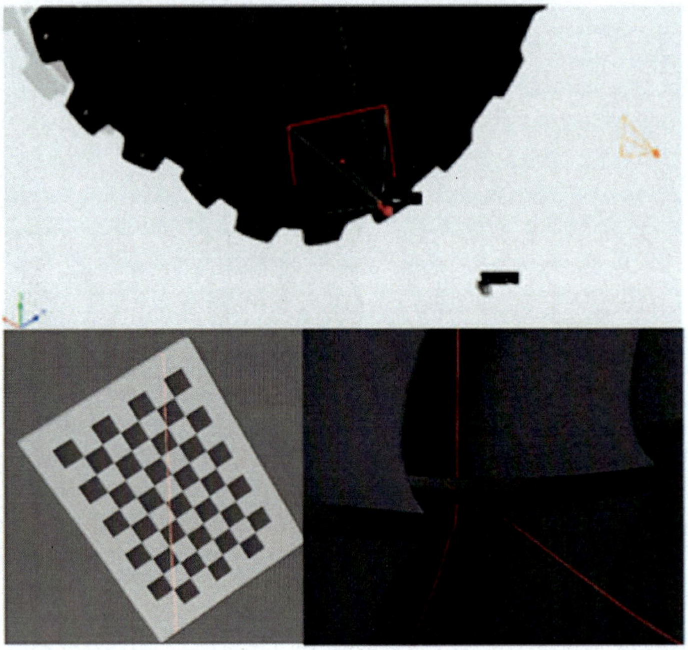

Fig. 8. Analog measurement system in software.

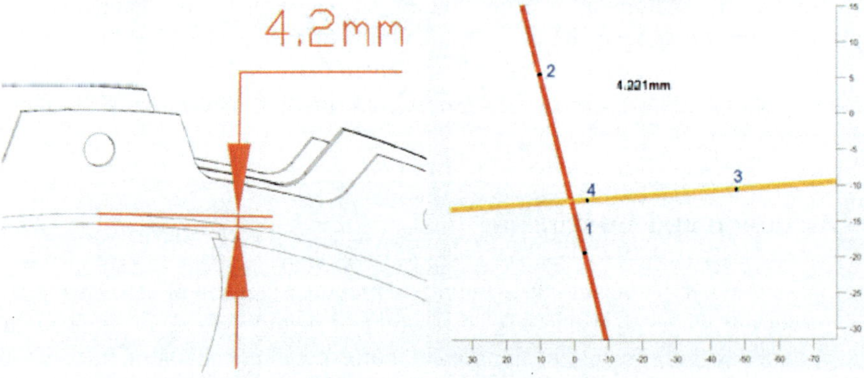

Fig. 9. Simulation experiment results.

by industrial cameras, in the future, the rotor tip clearance, stator casing roundness, rotor runout and other parameters could be measured simultaneously, which could provide more diversified data for aero engine assembly inspection.

Acknowledgements. The authors would like to acknowledge the financial support from the National Science and Technology Major Project of the Ministry of Science and Technology of China, China (Grant nos. 2017-VII-0010-0104).

References

1. Satish TN, Murthy R, Singh AK (2014) Analysis of uncertainties in measurement of rotor blade tip clearance in gas turbine engine under dynamic condition. Proc Inst Mech Eng Part G: J Aerosp Eng 228(5)
2. Sheard AG (2011) Blade by blade tip clearance measurement. Int J Rotat Mach 2011
3. Mevissen F, Meo M (2019) A review of NDT/structural health monitoring techniques for hot gas components in gas turbines. Sensors
4. Roeseler C, von Flotow A, Tappert P (2002) Monitoring blade passage in turbomachinery through the engine case (no holes). In: Proceedings, IEEE Aerospace conference, vol 6. IEEE
5. Vakhtin A, Chen S-J, Massick S (2009) Optical probe for monitoring blade tip clearance. In: 47th AIAA Aerospace sciences meeting including the new horizons forum and aerospace exposition
6. Neumann M et al (2015) A laser-optical sensor system for blade vibration detection of high-speed compressors. Mech Syst Signal Process 64
7. Violetti M et al (2012) New microwave sensing system for blade tip clearance measurement in gas turbines. Sensors 2012. IEEE
8. Stubbs JD, Muhammad AS (2020) Blade tip clearance measurement systems for high speed turbomachinery applications and the potential for blade tip timing applications. In: ASME Turbo Expo 2020: Turbomachinery technical conference and exposition. American Society of Mechanical Engineers Digital Collection
9. Han Y et al (2018) Online monitoring of dynamic tip clearance of turbine blades in high temperature environments. Meas Sci Technol 29(4)
10. Zhang J et al (2017) A blade tip timing method based on a microwave sensor. Sensors 17(5):1097
11. Wu J et al (2020) A novel blade tip clearance measurement method based on event capture technique. Mech Syst Signal Process 139
12. Jia, Sun (2013) Study on the measurement scheme of the coaxiality of the heavy-duty gas turbine horizontally against the start box structure and the rotating stator gap. Sci Technol Inf 12
13. Burge J, Duesler P (2003) Precise roundness and centrality control of thin shells. In: 44th AIAA/ASME/ASCE/AHS/ASC structures, structural dynamics, and materials conference
14. Song L et al (2013) Survey on camera calibration technique. In: 2013 5th International conference on intelligent human-machine systems and cybernetics, vol 2. IEEE

An Aero-Engine Assembly Deviation Analysis Method Based on Skin Model Shapes

Gang Zhao[1], Jinyue Li[1], Bing Zhang[2], Pengfei Zhang[1], Hui Wang[3], and Wenlei Xiao[1](✉)

[1] Beihang University, Beijing, China
ftd423@buaa.edu.cn
[2] AECC Shengyang Engine Research Institute, Shenyang, China
[3] Tsinghua University, Beijing, China

Abstract. In modern aero-engine assembly production, the assembly deviation analysis method plays an important role in product quality and efficiency. Various mathematical model based analysis methods are studied to realize assembly deviation prediction. However, most established analysis methods require abstraction and simplification of geometric deviations, which results in the inability to accurately characterizing the surface deviation features and analyzing the impact of deviation characteristics on the assembly. Therefore, a Skin Model Shapes based assembly deviation analysis method is proposed to enhance the accuracy and efficiency of deviation identification and assembly. This method enables the analysis of the assembly deviation caused by the surface deviation feature. The feature surface is represented by Skin Model Shapes, which is generated from surface measurement data. An assembly experiment of high pressure turbine components is performed to illustrate the procedure of this method, and the effectiveness of the method is verified by comparing the experimental and simulation results.

Keywords: Aero-engine · Stacking assembly · Virtual assembly · Skin Model Shapes

1 Introduction

The assembly process of the aero-engine largely determines its production efficiency and performance. Proper assembly instruction can improve the quality and efficiency of aero-engine assembly production. In modern aero-engine assembly production, assembly deviation analysis results are used for assembly evaluation and guidance. Thus, novel assembly deviation analysis methods are urgently needed to predict the effects of each part deviation on the status of the assembly. Relevant researches are mainly focused on the mathematical model used to describe the variation of the tolerance zone, and the accepted analytical method used to evaluate the effect of part deviations on the whole product, depends on the applicable conditions [1]. Various model-based analysis methods suitable for different conditions have been presented in the published literature. Desrochers et al. [2] introduced the topologically and technologically related surfaces (TTRS). In this

H. Ding (ed.), *Aerospace Mechatronics and Control Technology*, Springer Aerospace Technology,
https://doi.org/10.1007/978-981-16-6640-7_7

model, parts are represented by a successive binary tree formed by associated elementary surfaces such as cylinders, spheres, planes and helical, etc. Later, Desrochers et al. [3] further developed a Matrix model from the TTRS model. It used the homogeneous transform matrix, to describe the tolerance zones and the clearances, which is commonly used in robotic modeling. This method is also applicable to tolerance transmission and analysis in three dimensions. Chase and Gao et al. [4, 5] proposed the direct linearization method (DLM) for tolerance analysis of 2D/3D mechanical assemblies. This method uses the vector loops based model to represent the dimensional chains produced in the assembly. Dimensional deviation, geometric feature deviation, and deviation due to small kinematic adjustments in assembly are comprehensively considered. Desrochers et al. [6] combined the Jacobian model [7] that describes deviation transmission with transformation matrices and the Torsor model [8] that characterizes tolerance zones to introduce the unified Jacobian-Torsor model. The Jacobian based method enables the transformation of single parts deviations to the whole assembly [9]. Davidson et al. [10, 11] proposed the Tolerance-Map (T-Map) model. In the T-Map model, a hypothetical Euclidean point space is established, whose size and shape reflect the variational possibilities of the target feature. Some other methods, such as Feature based topologically and technologically related surfaces (FTTRS) [12] and Manufacturing-Map (M-Map) [13], are derived from the above methods.

However, the methods mentioned above are considered to be difficult to comprehensively consider the detail deviation information and the mutual influence on the connection surface between parts [14]. These model-based methods need to simplify featured surface to an ideal one (plane, cylinder, circle, etc.), whose rotation and translation matrixes are applied to deviation analysis. This leads to the difference between the analysis results and the actual assembly. With the development of computer technology, some researches about computer aided tolerancing have been done to solve this problem. Forslund et al. [15] used the 3D scanner data to evaluate turbine structure assemblies. Schleich et al. [16] proposed the concept of Skin Model Shapes. These methods use point clouds or surface meshes to represent the surface feature of a workpiece.

Based on the research status, this paper presents an aero-engine assembly deviation analysis method derived from Skin Model Shapes. The effect of surface deviation feature on rigid body assembly is considered. The mesh-represented model also brings the possibility of multi-physical environment assembly simulation.

The rest of this paper is structured as follows. In Sect. 2, the characteristics of aero-engine assembly are introduced. The concept of Skin Model Shapes is highlighted is Sect. 3. The simulation method for aero-engine assembly is presented in Sect. 4. A case study of rotor component assembly is given based on the simulation method. The conclusion of this work is given in the last part.

2 Aero-Engine Stacking Assembly

As the core component of an aero-engine, high pressure rotor puts forward high requirements of assembly. Therefore, stacking assembly technology is widely used in aero-engine rotor assembly to ensure product quality. This technology is realized by using the Part Measurement and Stacking Prediction System (SPS), as shown in Fig. 1.

Fig. 1. The part measurement and stacking prediction system.

This system is mainly composed of two parts: The air bearing table, including an air bearing, a self-aligning and tilting mechanism, an angular encoder and a chuck for holding workpiece. The measuring system, including the displacement sensors, adjustable clamps for sensors and an industrial PC for data acquisition.

The process of the rotor stacking assembly is shown in Fig. 2. After the SPS device starts up, each part is installed on the air bearing platform to be measured. During the measurement, the geometric parameters of the part need to be input firstly. Then, the air bearing platform is rotated gently. The measurement datum of the part is determined by two sensors, and the remaining sensors are used to measure the runout of the vital surface. Thereafter, the collected data is input into the built-in model of the SPS, and an optimized result is fed as assembly guidance. After the assembly, the new component is measured again, and the stacking optimization is performed with the newly measurement data as before. The above steps will be repeated until the assembly is completed.

Ideally, the assembly results should be consistent with the model optimization results. However, the prediction results of the SPS device are possible to be incompatible with the actual assembly results. A variety of reasons can lead to this result, such as parts deformation, deviation caused by the operation, unexpected clearance between interface and so on. After a preliminary analysis, the built-in analysis model of the SPS device might be derived from the Jacobian-Torsor method, which neglected the effect of surface deviation feature. Therefore, the research is focused on finding a method which can transfer the influence of contact surface deviation feature to the final assembly results.

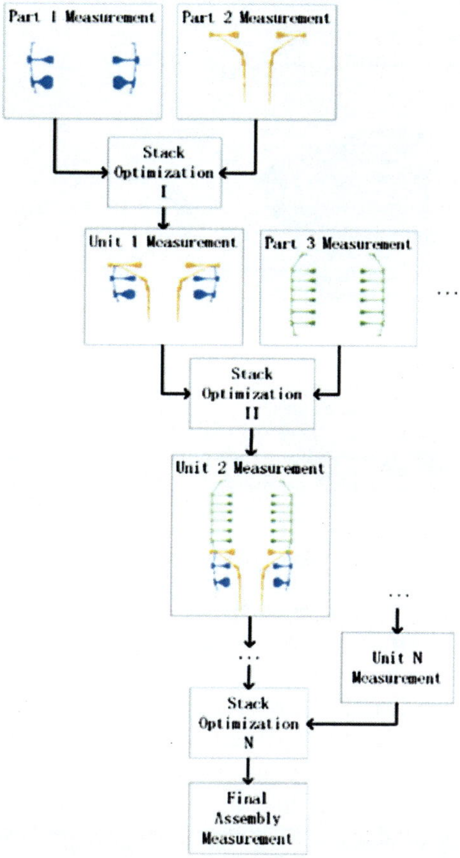

Fig. 2. High pressure rotor stacking process diagram.

3 Concept of Skin Model Shapes

Skin Model is an abstract model of the physical interface between a part and its environments, in contrast to other models for computer aided analysis make strict assumptions to the target surface. It comprises the deviations brought in by manufacturing and assembly processes [17]. However, the initial Skin Model is an infinite model, which is not suitable for computer analysis. Thus, a discrete Skin Model Shapes method is proposed for computer processing. The point clouds or surface meshes are used in Skin Model Shapes to represent the surface feature of a workpiece. The idea of Skin Model Shapes is to add deviation to the nominal model. The deviation could be generated by random theory, trained by samples, or directly observed [3]. It is suitable for various applications such as tolerancing, manufacturing, and inspection.

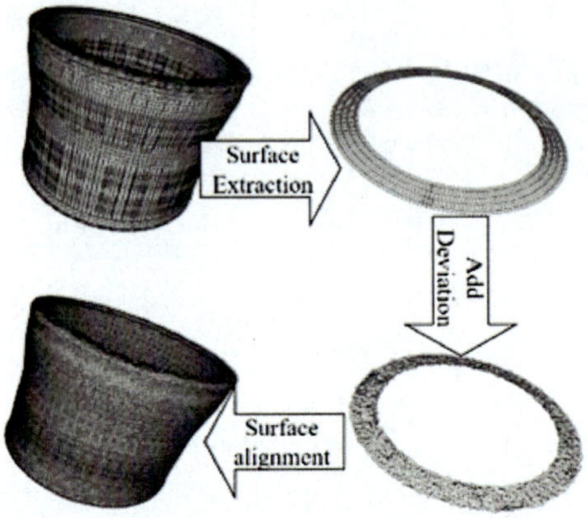

Fig. 3. Skin Model Shapes generation diagram.

Generally, the Skin Model Shapes based assembly deviation analysis method is consists of three steps as shown in Fig. 3: (1) the construction of Skin Model Shapes, (2) assembly simulations, and (3) characteristic extraction.

Firstly, 3D Skin Model Shapes is generated from the nominal model of a part. Then, the key surface is extracted from 3D model. The deviation is added to the extracted surface. At last, the surface with deviation information is returned to the nominal model. After completing the steps above, a Skin model Shape is ready for analysis. At present, Skin Mode Shapes based applications are mainly oriented to rigid bodies. However, it is possible to combine it with Finite Element Modelling to realize the Coupling analysis of mechanics and geometry.

4 Virtual Assembly Analysis

Base on the concept of Skin Model Shapes, virtual assembly analysis is proposed through three steps: First, the Skin Model Shapes of workpieces are constructed from measured data. Then, a virtual assembly is realized by a collision detection algorithm. At last, the characteristics of the final assembly are calculated.

4.1 Assembly Analysis Modelling

The model in this work is mainly consisting of two parts: the description of surface feature at the interface and the description of deviation transmission.

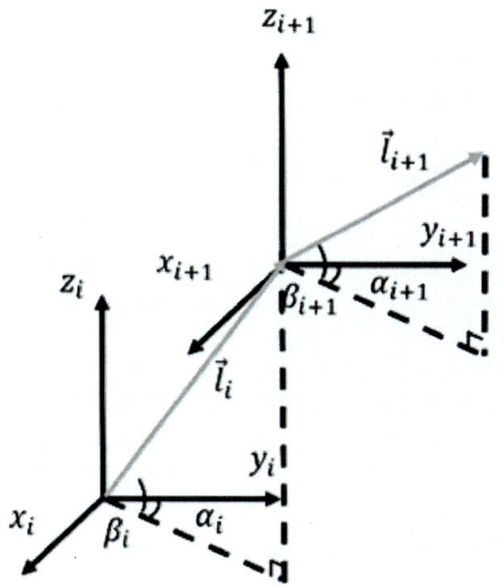

Fig. 4. Spatial dimensional chain and its parameters.

The description of deviation transmission is achieved by the spatial dimensional chain. All workpieces are assumed to be rigid bodies. Hence, the general spatial Dimensional Chain model can be expressed by:

$$\vec{l}_F = \vec{l}_1 + \vec{l}_2 + \ldots + \vec{l}_n \tag{1}$$

where the vector $\vec{l}_F$ represents the functional requirement of final assembly, $\vec{l}_i(i = 1, 2, \ldots, n)$ is the vector that represent the relative position of feature within a workpiece, the vector $\vec{l}_i$ is shown as:

$$\vec{l}_i = l_i \cdot [\cos\beta_i \ \cos\beta_i \ \sin\beta_i] \begin{bmatrix} \cos\alpha_i \\ \sin\alpha_i \\ 1 \end{bmatrix} \tag{2}$$

where l_i is the length of the unit in dimension chain, α_i and β_i are the two attitude angles of vector $\vec{l}_i$, as shown in Fig. 4.

In order to preserve the surface feature information as complete as possible, the surface is represented by the Skin Model Shapes generated directly from the measured data rather than a mathematical model. An ideal plane of each surface is generated for assembly simulation and characteristics calculation. But unlike the other model-based model, those ideal planes are used for deviation transmission analysis rather than contact deviation analysis, as shown in Fig. 5.

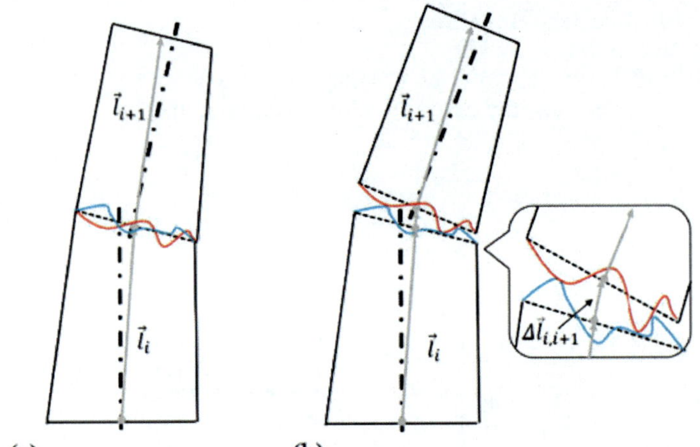

Fig. 5. Different between assembly analysis methods. **a** Ideal surface assembly. **b** Featured surface assembly

As depicted in the figure above, a new vector $\Delta\vec{l}_{i,i+1}$ is introduced into the dimensional chain. This is the displacement caused by non-ideal surface contact. Hence, the modified spatial dimensional chain is represented as:

$$\vec{l}_F = \vec{l}_1 + \Delta\vec{l}_{1,2} + \vec{l}_2 + \Delta\vec{l}_{2,3} \ldots \Delta\vec{l}_{n-1,n} + \vec{l}_n \tag{3}$$

where vector $\Delta\vec{l}_{i,i+1}(i = 1, 2, \ldots, n-1)$ is generated according to the Skin Model Shapes based assembly simulation results.

4.2 Contact Deviation Analysis Method

The model introduced in the last section requires the following inputs: the length of each workpiece in the dimensional chain l_i, the attitude angle α_i and β_i and the vector $\Delta\vec{l}_{i,i+1}$ represents the interface state. l_i can be obtained by measuring the geometric feature of the workpiece, while attitude angle α_i and β_i can be obtained by measuring concentricity. However, the contact deviation vector $\Delta\vec{l}_{i,i+1}$ is unmeasurable. Thus, a contact deviation analysis method is needed to calculate $\Delta\vec{l}_{i,i+1}$.

This problem is solved by using collision detection algorithm in Unity3D (Fig. 6). The generated Skin Model Shapes must be converted to mesh form. Then, mesh model are imported into Unity3D software. Key components including the MeshCollider and Rigidbody are added to the 3D Model to perform contact analysis. The contact deviation vector $\Delta\vec{l}_{i,i+1}$ is then calculated according to the mesh information of Skin Model Shapes.

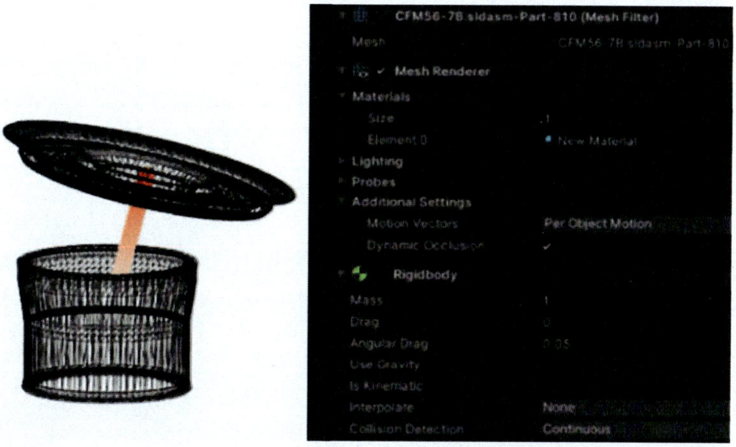

Fig. 6. Contact analysis in Unity, key components meshes (in left) used for analysis and calculated vector (in red)

5 Simulation and Assembly Test

The proposed analysis method is verified by the stacking assembly of the aero-engine test component. These test components including two workpieces: the high pressure turbine disk and the high pressure turbine drum. The 3D model of the component is shown in Fig. 7.

Fig. 7. 3D model of the test component

Firstly, the geometric data of the workpieces is measured by trilinear coordinates measuring instrument. In this experiment, the heights between two end surfaces of workpieces are essential data to establish the spatial dimension chain. Thus, the vertical distance between two installation flanges of high pressure turbine disk and that between two installation flanges of high pressure turbine drum are measured. The measured geometric data is shown in Table 1.

Table 1. The geometric data of test pieces

No	Part name	Geometric characteristics	Unit	Values
1	HPT* disk	Height of top and bottom flanges	mm	116.09
2	HPT drum	Height of top and bottom flanges	mm	262.71

*HPT: High pressure turbine

Thereafter, a reference coordinate is established to describe the spatial dimension chain.

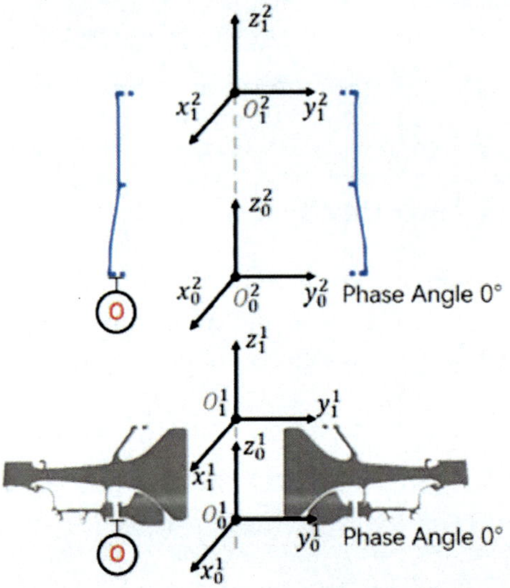

Fig. 8. Spatial relationship of different coordinates

In the process of the experiment, by using the SPS Device, the key surface run out of every single part is measured. Then, geometric tolerance is calculated through the measured data. Two conjoined coordinates are established for each part. Coordinate O_0^i is attached to the bottom surface, which is also the datum, the origin point locates at the center of the bottom circle. The axis x points to the position of 0° phase angle. The axis z points from O_0^i to O_1^i. The axis y is determined by the right-hand rule. O_1^i is attached to the top surface. The origin point locates at the center of the top circle, and the determination of the axes is the same. Hence, the length of each vector l_i is determined by measured geometric data and the vector $\vec{l}_i$ points from O_0^i origin to O_1^i origin. The spatial relationship of coordinates is shown in Fig. 8.

The spatial relationship between O_0^i origin and O_1^i origin is inferred from the tolerance information, which can be calculated by the SPS Device. Thus, the next step of the experiment is to measure the surface runout of the workpiece. The measurement results are shown in Table 2.

Table 2. Measurement of workpiece geometric tolerance

Measured items	Concentricity		Parallelism		Datum
	Value (mm)	Phase angle (°)	Value (mm)	Phase angle (°)	
HPT disk front flange axial runout	\	\	0.0058	110	HPT disk rear flange
HPT disk front flange radial runout	0.0049	30	\	\	HPT disk rear flange
HPT Drum front flange axial runout	\	\	0.0113	256	HPT drum rear flange
HPT Drum from flange radial runout	0.0057	42	\	\	HPT drum rear flange

Thereafter, Skin Model Shapes of key surfaces are generated from the measured runout data, as is shown in Fig. 9. Since the existing equipment is a point contact displacement sensor, repeated measurements are implemented to generate a surface. After the Skin Model Shapes are generated, contact deviation analysis is performed by the developed Unity3D program.

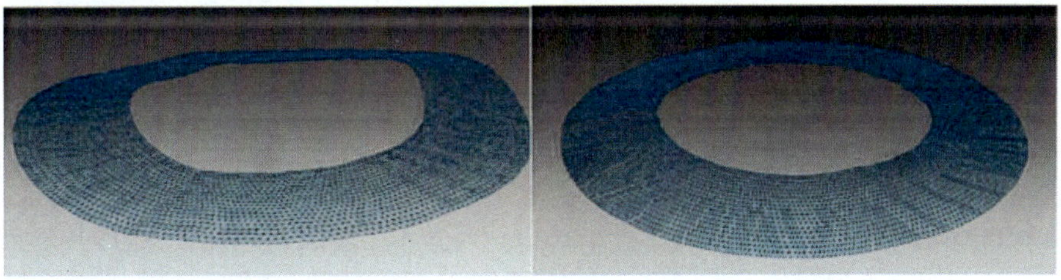

(a) (b)

Fig. 9. Skin Model Shapes generated from measurement data. **a** Skin Model Shapes of drum end surface. **b** Skin Model Shapes of HPT disk end surface

After the deviation vector is obtained by contact deviation analysis, a complete numerical assembly analysis can be carried out. According to the optimal assembly phase angle predicted by the SPS device, a workpiece assembly simulation is carried out. Then, workpieces are assembled and measured. The prediction of the SPS device, the prediction of the proposed method, and the measurement results are shown in Table 3.

It can be seen from Table 3 that the prediction result of the proposed method is 30% more accurate in phase angle, and 55% more accurate in concentricity. However, there is still a difference between the prediction and the measurement result. This may

Table 3. Comparison of different prediction results and the assembly results

Measured items	Measured result	SPS prediction	Model prediction
Concentricity, value, drum to disk	0.0355 mm	0.0101 mm	0.0240 mm
Concentricity, phase angle, drum to disk	203°	180°	219

be caused by many factors, such as inaccurate measurement, elastic deformation of the workpiece, internal stress between parts caused by the assembly, and so on. Future works in these aspects are designed to achieve a more accurate prediction model under the multi-physical environment.

6 Conclusion

Aiming at the stacking assembly process of aero-engine rotor components, this paper presents a new virtual assembly method combining 3D dimension chain and Skin Model Shapes. To improve the prediction accuracy of aero-engine stacking assembly, the influence of surface feature deviation on assembly accuracy is analyzed by Skin Model Shapes, and the influence of geometric deviation transfer between parts is modelled by spatial dimension chain. The effectiveness of the proposed method is validated by numerical simulations and assembly experiments. Results show that the simulation accuracy can be effectively improved by considering the influence of surface feature deviation on assembly accuracy.

Acknowledgements. The authors would like to acknowledge the financial support from the National Science and Technology Major Project of the Ministry of Science and Technology of China, China (Grant Nos. 2017-VII-0010-0104).

References

1. Cao Y, Liu T, Yang J (2018) A comprehensive review of tolerance analysis models. Int J Adv Manuf Technol 97:3055–3085
2. Desrochers A, Clement A (1994) A dimensioning and tolerancing assistance model for CAD/CAM Systems. Int J Adv Manuf Technol 9(6):352–361
3. Desrochers A, Riviere A (1997) A Matrix approach to the representation of tolerance zones and clearances. Int J Adv Manuf Technol 13(9):630–636
4. Chase KW, Gao J, Magleby SP (1995) General 2-D analysis of mechanical assemblies with small kinematic adjustments. J Des Manuf 5:263–390
5. Gao J, Chase KW, Magleby SP (1998) Generalized 3-D tolerance analysis of mechanical assemblies with small kinematic adjustments. IIE Trans 30:367–393
6. Desrochers A, Laperriere L, Ghie W (2003) Application of a unified Jacobian-Torsor model for tolerance analysis. J Comput Inf Sci Eng 3(1):2–14
7. Laperriere L, Lafond P (1999) Modeling tolerances and dispersions of mechanical assemblies using virtual joints. In: Proceedings of ASME design engineering technical conferences, 12

8. Bourder P, Mathieu L, Lartigue C, et al (1996) The concept of the small displacement torsor in metrology. Ser Adv Math Appl Sci 40
9. Laperriere L, Lafond P (1999) Tolerance analysis and synthesis using virtual joints. In 6th CIRP international seminar on computer aided tolerancing, 45
10. Davidson JK, Shah JJ (2002) A new mathematical model for geometric tolerances as applied to round faces. J Mech Des 124(4):609–622
11. Mujezinovic A, Davidson JK, Shah JJ (2004) A new mathematical model for geometric tolerances as applied to polygonal faces. J Mech Des 126(3):504–518
12. Liu Y, Gao S, Wu Z et al (2003) Hierarchical representation model and its realization of tolerance based on feature. Chin J Mech Eng 39(3):649–650
13. Jiang K, Davidson JK, Shah JJ, et al (2013) Using tolerance-maps to transfer datum plane from design tolerancing to machining tolerancing. In Proceedings of the ASME 2013 international design engineerign technical conferences and computers and information in engineering conference
14. Ding S, Zhaeng X, Bao J, Zhang J (2021) A comprehensive study on deviation modeling of multistage revolving components assembly. J Donghua Univ (Nat Sci) 47(1):77–85
15. Anders F, Rikard S, Johan L, Alejandro VG (2011) Virtual robustness evaluation of turbine structure assemblies using 3D scanner data. In Proceedings of the ASME 2011 international mechanical engineering congress & exposition, 1
16. Schleich B, Anwer N, Mathieu L et al (2014) Skin model shapes: a new paradigm shift for geometric vatiations modelling in mechanical engineering. Comput Aided Des 50(3):1–15
17. Ballum A, Dantan JY, Mathieu L (2010) Geometric toleracing of products. Language of tolerancing: Geospelling, M. ISTE Ltd and John Wiley & Sons Inc

Parameters Online Identification and Prediction of Solid Rocket Motor

Zhenyu Jiang and Rundong Ma(✉)

College of Aerospace Science and Engineering, National University of Defense Technology, Changsha 410037, P.R. China
jiangzhenyu00@yeah.net, marundong97@163.com

Abstract. Considering the ground-to-flight difference in solid rocket motor and uncertainties on motor performance parameters at the shutdown point, this paper proposes a method for online identification and prediction of solid rocket motor performance. The calculation model of the nozzle throat diameter and the propellant burning rate are established based on the analysis of internal correlation between motor parameters. Then by means of the real-time measured data from the missile-borne sensor, throat diameter and burning rate can be calculated. After the model parameters of throat diameter and burning rate were identified by the recursive least square algorithm, the shutdown time and thrust-time characteristics can be predicted based on identification results and prediction model. The simulation results show that this method has high identification and prediction accuracy, and could provide a feasible solution for online identification and prediction of the performance parameters of solid rocket motor.

Keywords: Solid rocket motor · Online identification · Prediction

1 Introduction

Solid rocket motor (SRM) has been widely used in launching vehicle and missile. Because the uncertainty of flight environment and ground-to-flight difference would cause a degree of fragmentation in motor performance parameters, such as thrust, shutdown time, which would cause deviation in speed and position at the shutdown point. Hence, the ground-to-flight difference has aroused extensive attention from international scholars in order to refine the design and control [1].

A great deal of researches have been conducted on SRM flight performance. By studying different elements that influencing the flight performance of VEGA motor which developed by European Space Agency (ESA), the ablation of the nozzle throat would cause obvious change in thrust efficiency and combustion efficiency [2–4]. By means of numerical simulation, it found out that thrust and pressure could be influenced by throat ablation and burning rate, and the change of throat ablation and burning rate had a great influence on pressure and thrust [5, 6]. Through analysis of the influence on motor performance which caused by thermal protection & liner ejected mass and deposition mass, it pointed out that further research should be taken on the changing

H. Ding (ed.), *Aerospace Mechatronics and Control Technology*, Springer Aerospace Technology,
https://doi.org/10.1007/978-981-16-6640-7_8

low of deposition quality and throat diameter under flight overload condition [7, 8]; In terms of static firing tests, it found out that lateral flight overload would cause eccentric ablation of nozzle throat lining [9]; Based on apparent acceleration model and motor internal ballistic model, the identification of specific impulse could be realized [10, 11]; Based on static firing test, identification model of nozzle throat ablation that could reflect the throat diameter changing law was established, and it could provide help for the prediction of the motor internal ballistic performance [12]. The change of the throat diameter could be regarded as a grey-box system, and different model were established to describe the change of the throat diameter, then the model parameters could be identified [13]. By means of neural network, the complex nonlinear relationship between specific impulse and its influencing factors could be described, and the prediction level of SRM performance could also be improved [14].

The above researches show that the ablation of nozzle throat and the deviation of propellant burning rate in actual flight are important influencing factors for thrust deviation. In this paper, SRM motor performance calculation model is proposed based on measured data. Subsequently, online identification is performed to identify the model parameters based on calculation result. Finally, motor performance shutdown time and thrust-time data are predicted. This work could serve as a basis for the online trajectory planning and guidance of the aircraft in the boost stage and the subsequent flight stage.

2 The Strategy of Online Identification and Prediction

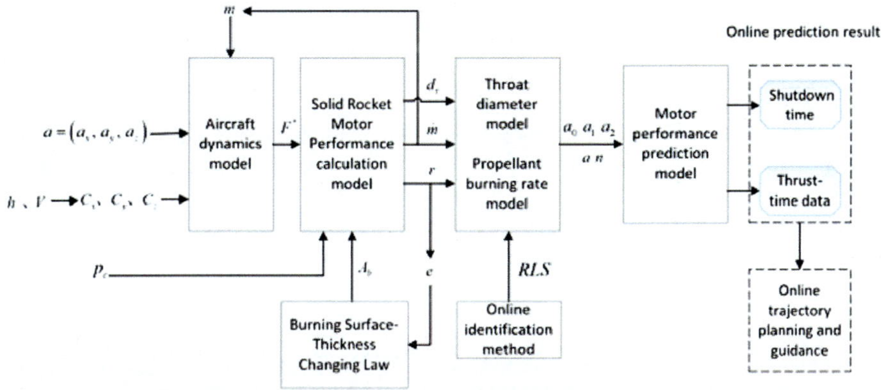

Fig. 1. Process of online identification and prediction of SRM parameters.

System identification is used to determine the mathematical model of the system through measurable input and output data [15]. If the system model structure has been determined, system identification could be simplified to the problem of model parameter estimation, namely parameter identification, which involves the mathematical model of the system, the input and output data of the observed system, and parameter estimation methods.

In order to realize parameter identification of SRM, it first needs to establish the parameter model to be identified. However, the relationship between the motor performance parameters, such as the shutdown time, thrust, mass flux, and throat diameter are related to each other, so the online identification of motor performance parameters is a dynamic iterative process. Then by means of establishing appropriate aircraft dynamics model, such as motor performance calculation model, parameter model that need to be identified, and prediction model, the performance parameters could be estimated and updated, based on real-time measured information in actual flight, and the internal ballistic performance of the motor could be predicted.

In order to simplify the model and make it suitable for online identification, the following four assumptions are proposed [12, 16]:

(a) C^* denotes propellant characteristic velocity, it is known and remains unchanged during the identification process.
(b) Compare to the thrust, the aerodynamic force in the booster stage is a small amount. Therefore, the aerodynamic coefficient has no deviation in actual flight.
(c) The burning surface-wall thickness changing law obeys the design.
(d) The cross-sectional area of the nozzle outlet A_e is a constant value. Based on the above assumptions, our identification and prediction strategy is shown in Fig. 1.

3 Online Identification and Prediction Method

3.1 SRM Performance Calculation Model

According to the real-time apparent acceleration of the aircraft measured by the missile-borne sensors and the real-time updated aircraft mass m and aerodynamic drag X, the current thrust F^* can be calculated through the aircraft dynamics model [3].

$$F^* = ma + X \tag{1}$$

$$m = m_0 - \int_0^{ta} \frac{P_C A_t}{C^*} dt - \int_0^{ta} m_{in} dt \tag{2}$$

$$X = \frac{1}{2} \rho V^2 C_X S \tag{3}$$

In the above formulas, m_0 is the take-off weight of the aircraft, t is the current time, p_c is the pressure of the combustion chamber, m_{in} is the mass flux corresponding to the negative mass of the engine, and is the known quantity determined through ground tests, ρ is the atmospheric density, V is the current flight speed, C_X is the aerodynamic drag coefficient, and S is the reference area of the aircraft.

Take the initial throat area as the initial value of the throat area iteration $A_{t,i}$, then the corresponding mass flux $\dot{m}_i$ can be obtained by Eq. (4).

$$\dot{m} = \frac{p_c A_t}{C^*} \tag{4}$$

Since A_e is constant value during the flight, the value of A_e/A_t can be obtained. Then through equation p_e/p_ccan be obtained.

$$\frac{A_e}{A_t} = \frac{\Gamma}{\left(\frac{p_e}{p_c}\right)^{\frac{1}{k}} \sqrt{\frac{2k}{k-1}\left[1 - \left(\frac{p_e}{p_c}\right)^{\frac{k-1}{k}}\right]}} \tag{5}$$

In the formula, the specific heat ratio k and Γ of gas are obtained through thermodynamic calculation.

Bring p_e/p_c into the following Eq. (6), the exhaust speed and vacuum exhaust speed can be calculated.

$$\begin{cases} u_{eo} = \sqrt{\frac{2k}{k-1} RT_f \left[1 - \left(\frac{p_e}{p_c}\right)^{\frac{k-1}{k}}\right]} \\ u_{efo} = u_{eo} + \frac{A_e p_e}{C_D A_{t,i} p_c} \end{cases} \tag{6}$$

In the formula, $C_D = 1/C^*$.

After obtaining the vacuum exhaust velocity u_{efo} and the mass flux $\dot{m}_{\mathrm{i}}$, the corresponding thrust can be calculated by Eq. (7).

$$F_i = \dot{m} \cdot u_{efo} - A_e P_a \tag{7}$$

In the formula, P_a is the ambient atmospheric pressure, which is calculated by the atmospheric parameter model and the altitude of the aircraft.

The function relation between thrust, mass flux and throat area $F_i = f(\dot{m} A_t)$ is established.

After obtaining the actual thrust F^*, construct the iteration format as follows:

$$\begin{cases} \varepsilon = At \\ \varepsilon_{i+1} = \varepsilon_i + (F^* - F_i) \frac{\Delta\varepsilon}{(F' - F_i)} \\ F_i = f(\dot{m}, \varepsilon_i) \\ F' = f(\dot{m}, \varepsilon_i + \Delta\varepsilon) \end{cases} \tag{8}$$

Set $\Delta\varepsilon = 1\mathrm{e}{-8}$ m, and the termination condition of iteration is that the error between F^* and F_i is less than 0.1 N.

With the calculated throat area A_t and mass flux $\dot{m}$, the throat diameter d and burning rate r can be calculated by the follow formulas.

$$d = \sqrt{\frac{4A_t}{\pi}} \tag{9}$$

$$r = \dot{\mathrm{m}}_b / \rho_p A_b \tag{10}$$

In the formula, ρ_p is the propellant density, and A_b is the burning surface calculated by the known burning surface—thickness law.

3.2 Throat Diameter Model and Propellant Burning Rate Model

The mechanism of throat ablation is related to the propellant components and the type of throat liner. Although the influencing factors are complex, but most of the changes in throat diameter over time can be described by polynomial models [12, 13]. For studied motor in this paper, the throat ablation law obtained from ground tests approximately satisfies the linear ablation law, so this article uses a quadratic polynomial to describe throat diameter d:

$$\mathrm{r}(t) = a_0 + a_1(t - t_1) + a_2(t - t_1)^2 \tag{11}$$

In the formula, a_0 is the coefficient of the zero degree term, a_1 is the coefficient of the first order term, and a_2 is the coefficient of the quadratic term, t_1 is the time when the throat diameter begins to change, t is the current moment.

Regardless of the influence of erosive burning, the burning rate model adopts the exponential model in Eq. (12).

$$r = ap_c^n \tag{12}$$

In the formula, a is the burning rate coefficient, and n is the burning rate pressure index.

Because the coefficients in the burning rate model need to be identified online, the model needs to be linearized. A convenient form can be obtained by taking the logarithm of both sides of Eq. (12)

$$\ln(r) = \ln(a) + n\ln(p_c) \tag{13}$$

Based on the selected burning rate and throat diameter models, it is considered that the order and structure of the throat diameter and burning rate models remain unchanged during flight, while the model parameters would change. Therefore, the five coefficients of a_0, a_1, a_2, a, n need to be identified online.

3.3 Online Identification Method

Commonly used online identification algorithms include recursive least squares method (RLS), recursive maximum likelihood method, Kalman filter method and so on. The recursive maximum likelihood method can identify both linear and nonlinear models, while the recursive least square method is suitable for linear models. Considering that the throat diameter and burning rate models used in this article can be converted into linear models, so the recursive least squares algorithm is used for SRM parameters identification. The principle of this method is as follows:

Suppose the model of the parameters to be sought is

$$y(j) = h_\xi(j)\xi + \upsilon(j) \tag{14}$$

In the formula, ξ is the parameter vector to be estimated. In this section, $\hat{\xi}_j$ is used to refer to the parameter estimated value obtained by the observation data before time j,

$y(j)$ is the observation vector, $\upsilon(j)$ is the observation noise, and $h_\xi(j)$ is the observation matrix. Then the total equation composed of time j and the observation equation before time j is

$$Y(j) = H(j)\xi + V_\xi(j) \tag{15}$$

So there is

$$Y(j+1) = \begin{bmatrix} Y(j) \\ y(j+1) \end{bmatrix} \tag{16}$$

$$H(j+1) = \begin{bmatrix} H(j) \\ h_\xi(j+1) \end{bmatrix} \tag{17}$$

$$H^T(j+1)H(j+1) = \begin{bmatrix} H(j) \\ h_\xi(j+1) \end{bmatrix}^T \begin{bmatrix} H(j) \\ h_\xi(j+1) \end{bmatrix} \tag{18}$$

Define the information matrix $B = H^T H$, note that $P = B^{-1}$, according to the matrix inversion formula

$$P(j+1) = P(j) - K(j)h_\xi(j+1)P(j) \tag{19}$$

Among them, the gain matrix

$$K(j) = P(j)h_\xi^T(j+1)\left[I + h_\xi(j+1)P(j)h_\xi^T(j+1)\right]^{-1} \tag{20}$$

Combined with the least squares estimation criterion

$$\begin{aligned} \hat{\xi}_j &= \left[H^T(j)H(j)\right]^{-1} H^T(j)Y(j) \\ \hat{\xi}_{j+1} &= \left[H^T(j+1)H(j+1)\right]^{-1} H^T(j+1)Y(j+1) \end{aligned} \tag{21}$$

Derive the parameter estimation in the recursive form:

$$\hat{\xi}_{j+1} = \hat{\xi}_j + K(j)\left[y(j+1) - h_\xi(j+1)\hat{\xi}_j\right] \tag{22}$$

Equations (20), (21), (23) constitute the calculation equations of the recursive least square method.

3.4 Motor Performance Prediction Model

The motor performance prediction model is the motor interior ballistic model which based on online identification results to update motor parameters. With the identification results, the prediction model can realize the prediction of the motor shutdown time and the thrust-time data.

$$p_c = \left(a\rho C^* A_b/A_t\right)^{1/(1-n)} \tag{23}$$

$$F = C_F p_c A_t \tag{24}$$

$$C_F = \Gamma\sqrt{\frac{2k}{(k-1)}\left[1-\left(\frac{p_e}{p_c}\right)^{\frac{k-1}{k}}\right]}+\frac{A_e}{A_t}\left(\frac{p_e}{p_c}-\frac{p_a}{p_c}\right) \tag{25}$$

In the above equations, A_t and r are both updated by online identification. The definition of each parameter in the formula can be found in reference [11].

4 Numerical Simulation

4.1 Example Design

Taking a single-chamber dual-thrust solid rocket motor as an example to verify the feasibility of online identification and prediction of motor performance parameters. By comparing the deviation between thrust-time data under design conditions and online prediction data, evaluate the performance of online identification and prediction methods. To acquire the actual measured value of the sensor in actual flight, using the simulation method that adding the measurement error on the acceleration and pressure data obtained by the standard ballistic simulation to construct the measured data.

The motor throat liner adopts C/C composite material, and the propellant adopts a three-component composite solid propellant. The parameters in the throat diameter model and the propellant burning rate model are taken respectively $a_0 = 0.04$, $a_1 = 3.0853\text{e} - 4$, $a_2 = -1.0992\text{e} - 5$, $a = 0.010325$, $n = 0.25$ as the design reference value.

Suppose the sampling frequency of the measuring sensor is 200 Hz, the range of the pressure measuring sensor is 0-10 MPa, and the acceleration measurement range is 0–10 g. In order to test the influence of the sensor accuracy on the identification result, three simulation verification examples were designed by combining different accuracy of the sensor, and the indicators are shown in Table 1.

Table 1. Combination of sensor accuracy.

Example	Pressure (FS) (%)	Acceleration (FS) (%)
1	0.05	0.1
2	0.05	0.5
3	0.1	0.5

4.2 Result and Analysis

Based on the throat diameter and burning rate calculated online, the recursive least squares algorithm is used to identify a_0, a_1, a_2, a, n in the throat diameter and burning rate model. The identification results of the model parameters are shown in Figs. 2, 3, 4, 5, and 6. The identification results of most parameters converge to the reference value within 3s. The change of measurement accuracy has a obvious influence to the identification results. Among these examples, the combination example 1 with the highest measurement accuracy has the highest identification accuracy, and the combination example 3 with the lowest measurement accuracy has the worst identification accuracy. The results of parameter identification accuracy under different measurement accuracy are shown in Table 2.

Table 2. Parameter identification accuracy.

Parameter	Example 1 (%)	Example 2 (%)	Example 3 (%)
a_0	0.002	0.005	0.006
a_1	0.8	2.5	3.5
a_2	0.15	0.2	5
a	0.4	1.5	0.9
n	0.21	0.30	0.33

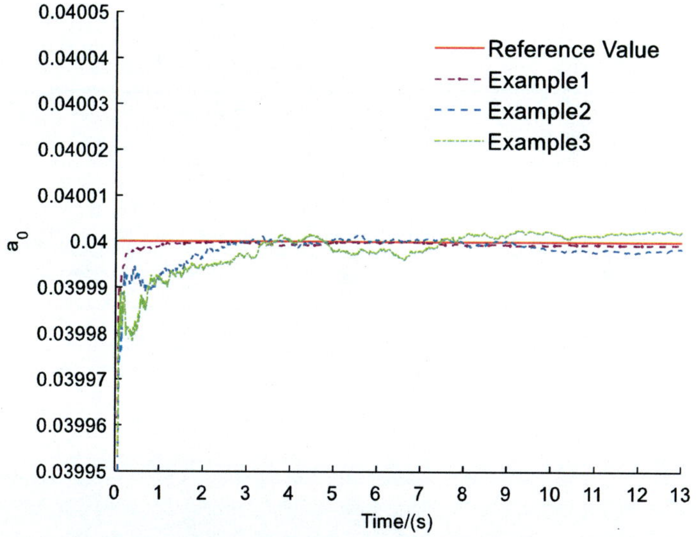

Fig. 2. Identification curve of a_0.

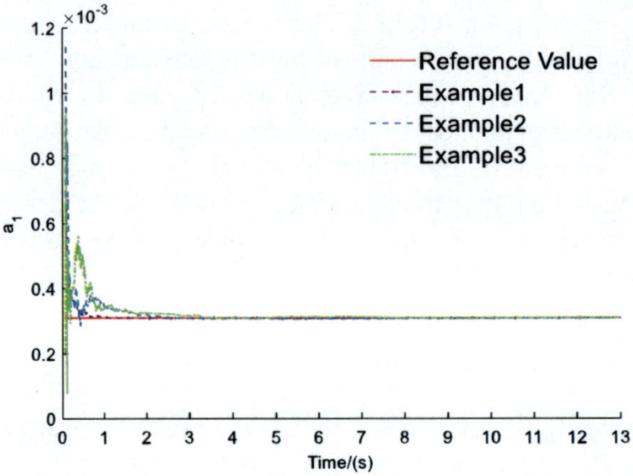

Fig. 3. Identification curve of a_1.

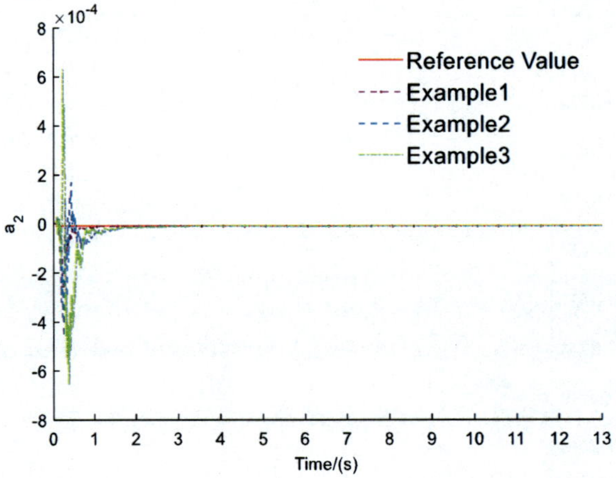

Fig. 4. Identification curve of a_2.

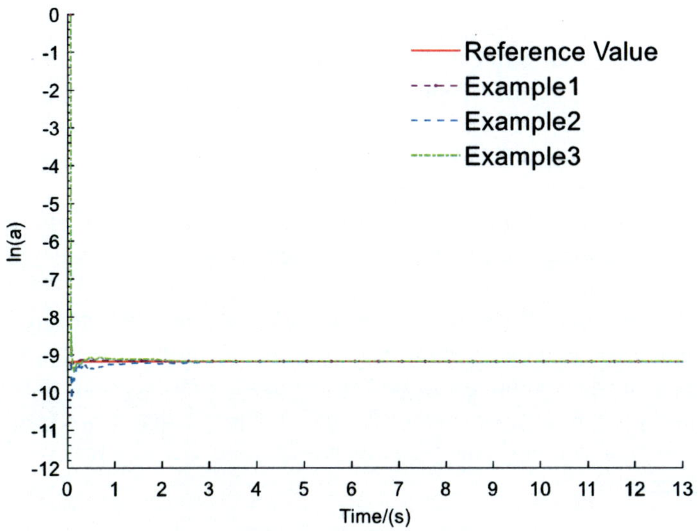

Fig. 5. Identification curve of a.

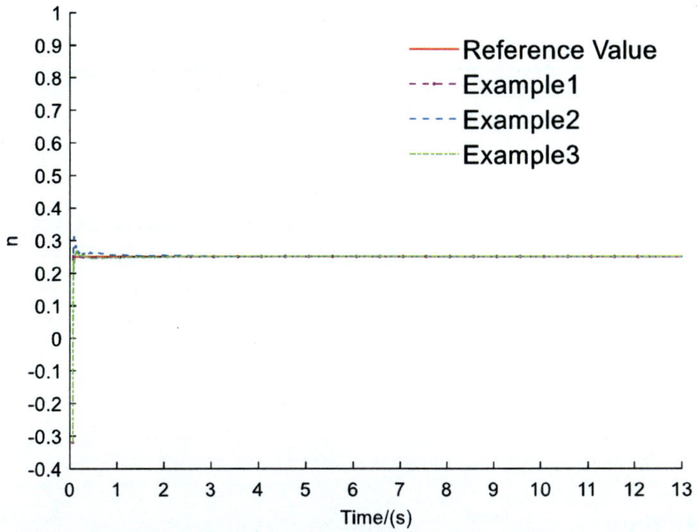

Fig. 6. Identification curve of n.

After obtaining the parameters of the throat diameter model and the propellant burning rate model, by means of the motor performance prediction model, the performance parameters can be predicted. Adopt the identification result of example 3, the prediction result of motor performance is shown in Fig. 7, t_p represents the time when the prediction starts, and based on the identified parameters acquired at this time, the motor performance is predicted in the remaining time.

Figure 7 implies that when the predictions are taken from the 1 s and 2 s, the motor performance prediction exists a certain degree of deviation on shutdown time and thrust-time data, and when the prediction starts from 3 s, the prediction result is more consistent with the reference value. The reason is that the values of the parameters to be identified directly affect the prediction accuracy, and the online identification algorithm needs to accumulate certain data before the identification results converge to the reference value. As shown in Figs. 2, 3, 4, 5, and 6, the identification parameters converged after 3 s, so when the prediction started from this moment, the motor performance prediction value has a higher accuracy.

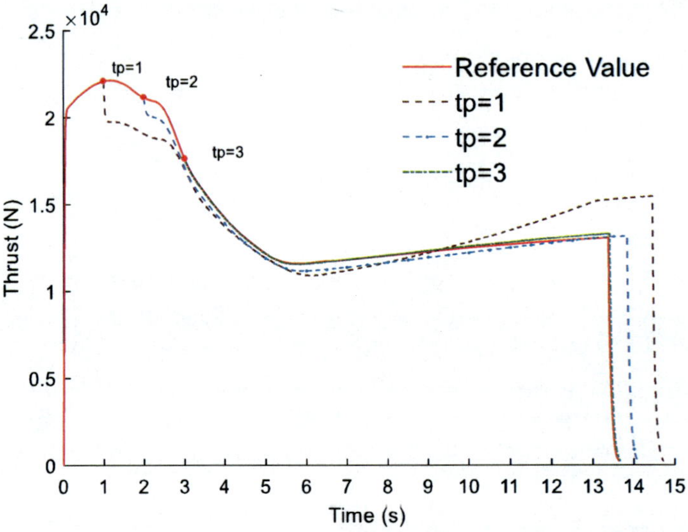

Fig. 7. Prediction of motor performance.

5 Conclusion

In order to address the problem of ground-to-flight difference in motor performance might caused by throat ablation and change in burning rate, this paper proposes a method of online identification and prediction for SRM parameters. Based on the online measured data, throat diameter and propellant burning rate can be calculated. By means of constructing the model of throat diameter and propellant burning rate, the model parameters are identified, then the prediction of motor performance is realized, based on the identification results. The simulation results show that:

(a) The proposed online identification and prediction strategy can complete the online identification of motor throat diameter and burning rate model parameters, could realize the online prediction of thrust-time performance, and it would also provide support for online trajectory planning and guidance in the boost phase and subsequent flight phases.
(b) The accuracy of the measuring sensor has an important influence on the accuracy of online identification and prediction. To ensure the effectiveness of online identification and prediction, the accuracy of acceleration and motor combustion chamber pressure measurement should be improved as much as possible.

The method in this paper provides a feasible solution for online identification and prediction of the performance parameters of SRM. However, withnot considering the influence of aerodynamic deviation and thrust deflection on identification and prediction, the related motor performance model is simplified to a certain extent. So these researches will be carried out in the future.

Acknowledgements. First of all, I would like to thank scholars who are engaged in the field of SRM. Without their research, I would not have the idea of this article. Secondly, I would like to thank my colleagues and students for their help in completing this article. It is their sincere help and constructive guidance that promoted the progress of the research.

References

1. Lv X, He G, Liu P, et al (2019) Requests and suggests for basic research on SRM combustion and flow. J Astronaut 40(10):59–68(Chinese)
2. Cavallini E, Favini B, Neri A (2014) Analysis and performance reconstruction of VEGA solid rocket motors qualification flights. AIAA. 2014–3805
3. Cavallini E, Favini B, Giacinto MD, et al (2012) Analysis of VEGA solid stages static firing tests towards the maiden Flight. In The 48th joint propulsion conference
4. Giliberti F, Angelone M, Fiorillo M, et al (2013) VEGA solid rocket motors flight performances results. In: The 49th joint propulsion conference
5. Bianchi D, Favini B, Neri A (2018) Performance reconstruction of VEGA solid rocket motors: VERTA flights experience. AIAA 2018–4692
6. Bianchi D, Grossi M, Favini B, et al (2020) Static firing ballistic reconstruction modelling and performance extrapolation to flight in SRMs. In: AIAA Propulsion and Energy 2020 Forum
7. Zhao J, Gong Y (2018) Flight performance analysis and reconstruction of SRM. J Solid Rocket Technol 41(01):7–10+27(Chinese)
8. Zhao J, Yu Z, Gong Y (2017) Discussion on ground-to-flight performance difference of SRM. J Solid Rocket Technol 40(02):135–140(Chinese)
9. Liu Z, Gao W, Zhang F (2018) Influence of acceleration on internal ballistics of solid rocket motor. J Solid Rocket Technol 41(02):169–175(Chinese)
10. He J, Hou X (2019) Analysis on specific impulse difference for SRM under ground and flight conditions. J Solid Rocket Technol 42(01):16–20(Chinese)
11. He J, Yu Z, Sun L, et al (2004) Internal ballistic calculation of solid rocket motors. J Solid Rocket Technol (01):20–21+45(Chinese)
12. Li Y, Zhou Y, Sun Z, et al (2020) Discrimination technology for the nozzle throat diameter erosion performance of solid rocket motor. J Projectiles, Rockets, Missiles Guidance. 40(02):60–62(Chinese)

Aerospace and Satellite Engineering

A Multi-missile Coverage Interception Strategy

Bao Song, Jianqiao Yu(✉), Xi Chen, Kang Niu, and Ziyuan Li

School of Aerospace Engineering, Beijing Institute of Technology, Beijing 100086, People's Republic of China
3120180077@bit.edu.cn

Abstract. Due to the error of target indication, the existing laser beam riding guidance weapon system is difficult to intercept the long-range target accurately. In order to solve this problem, a multi-missile coverage interception strategy is proposed in this paper: the total number of missiles determined is decomposed into a finite batch, and the interception probability is improved by adjusting the position of the same group of missiles in the interception plane. In addition, this paper proposes an Improved Grey Wolf Optimizer (IGWO). Based on the traditional Grey Wolf Optimizer, this algorithm generates the initial population uniformly by Hammersley sequence, and adopts a nonlinear convergence factor adjustment method, which can quickly and effectively optimize the location of the missiles. The simulation results show that the improved grey wolf optimizer improves the search accuracy, stability and convergence speed greatly. The multi-missile coverage interception strategy based on the improved grey wolf optimizer can effectively improve the interception accuracy.

Keywords: Multi-missile coverage interception strategy · Improved grey wolf optimizer · Hammersley sequence

1 Introduction

The modern battlefield environment becomes more and more complex, many countries in the world attach more and more importance to long-range precision strike weapons [1]. In many types of guided munitions, the guidance technology mainly includes autonomous guidance technology, remote control guidance technology, homing guidance technology and compound guidance technology. Remote guidance is a kind of guidance technology that sends guidance information to the ammunition through the guidance station to guide the ammunition to the target or the predetermined area [2]. Among all kinds of guidance technology, Optical (laser) beam riding guidance is an important way of remote control guidance. Under the continuous promotion of the development of laser technology, China has been equipped with a large number of laser beam guided ammunition.

Beam riding guided weapon system is widely used because of its strong anti-jamming ability, simple structure and low cost [3]. After the missile is launched, the guidance station generates the guidance beam with the center line of the beam aligned with the target. The missile senses its relative position in the guidance beam by the equipment on the missile, generates control command to guide the missile to fly along the center

H. Ding (ed.), *Aerospace Mechatronics and Control Technology*, Springer Aerospace Technology,
https://doi.org/10.1007/978-981-16-6640-7_9

line of the beam, and finally hits the target. However, the existing beam riding guidance system is difficult to intercept the long-range target accurately. The main reason for this phenomenon is that the guidance station has the target indication error in the process of controlling the guidance beam to track the target, and the error increases with the increase of the distance between the guidance station and the target. For example, when a missile intercepts a target 10 km away, an indication error of 0.5mil will result in a deviation of 5 m in the intercept plane, which greatly reduces the probability of the missile intercepting the target.

In order to solve this problem, this paper proposes a multi-missile coverage interception strategy under beam riding guidance system. This strategy divides the missiles into some groups. Meanwhile, an improved Wolf pack algorithm with high searching accuracy, good stability and fast convergence is proposed to optimize the intersection positions of the same group of missiles. The interception accuracy of the beam steering guidance system against the long-range target is greatly improved.

2 Multi-missile Coverage Interception Strategy

2.1 Brief Description of the Strategy

When a ground weapon system intercept s an important target, it usually needs to launch multiple missiles to increase the interception probability. If the target has no damage accumulation and each batch launches 1 missile with the same probability, then the probability of missiles can be expressed as

$$P_n = 1 - (1 - P_1)^n \tag{1}$$

It can be seen from Fig. 1 that the increase of interception probability is not directly proportional to the number of missiles consumed. It is inefficient to increase the intercept probability simply by increasing the number of missiles consumed.

Therefore, this paper proposes a multi-missile coverage interception strategy, which mainly includes:

(1) Decompose the determined number of missiles into finite groups.

$$n = n_1 + n_2 + \cdots n_m \tag{2}$$

(2) Adjust the intersection position of the same group of missiles and calculate the intercept probability P_{n_i}.
(3) Calculate the overall intercept probability.

$$P_n = 1 - \prod_{i=1}^{n} \left(1 - P_{n_i}\right) \tag{3}$$

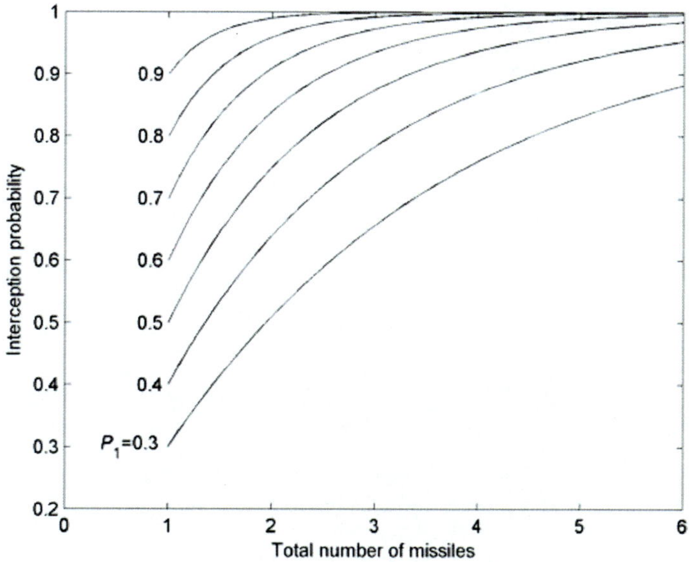

Fig. 1. Interception probability of n missiles

2.2 Intercept Probability of Multiple Missiles in the Same Group

The experimental data and theoretical analysis show that under the influence of the indication error, the virtual target point pointed by the center line of the beam is normally distributed near the real target point [4]. In the intersection plane of missile and target, the real target point is taken as the coordinate origin, then the distribution density function of virtual target point is

$$f(y',z') = \frac{1}{2\pi\sigma_y\sigma_z\sqrt{1-r_{yz}^2}}\exp\left[-\frac{1}{2\left(1-r_{yz}^2\right)}\left(\frac{(y')^2}{\sigma_y^2} - \frac{2r_{yz}y'z'}{\sigma_y\sigma_z} + \frac{(z')^2}{\sigma_z^2}\right)\right] \quad (4)$$

where σ_y and σ_z are the respective root mean square differences of guidance error along y' axis and z' axis, and r_{yz} is the correlation coefficient of the linear correlation degree of the two-dimensional random variable of guidance error.

The guidance information of a missile mainly comes from the deviation between itself and the center line of the beam which can only point to the virtual target point. Therefore, in this paper, the coordinate origin is shifted to the virtual target point, then the distribution density function of the real target point position corresponds to (5).

$$f(y,z) = f(-y',-z') = \frac{1}{2\pi\sigma_y\sigma_z\sqrt{1-r_{yz}^2}}\exp\left[-\frac{1}{2\left(1-r_{yz}^2\right)}\left(\frac{y^2}{\sigma_y^2} - \frac{2r_{yz}yz}{\sigma_y\sigma_z} + \frac{z^2}{\sigma_z^2}\right)\right] \quad (5)$$

Set the killing range of the missile as R, then the rendezvous between multiple missiles in the same group and the target is shown in Fig. 2.

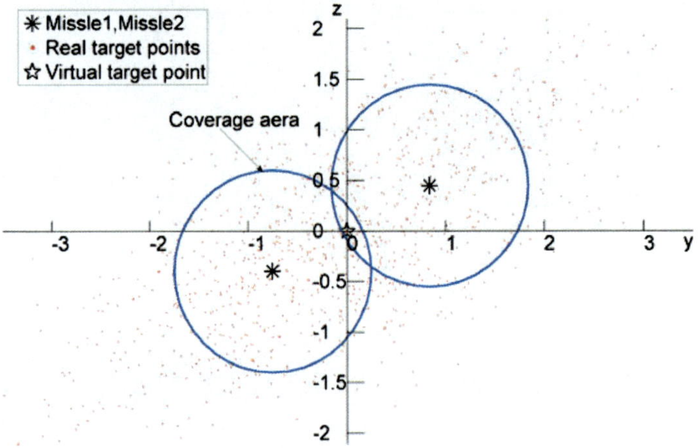

Fig. 2. Missile and target distribution

The missile interception probability can be expressed as

$$P = \iint_{\Omega} f(y, z)dydz \tag{6}$$

where Ω represents the impact coverage area of the missile, when the number of missiles in the same group is single

$$\Omega = \Omega_{m1} : (x - x_{m1})^2 + (y - y_{m1})^2 \leq R^2 \tag{7}$$

when there are multiple missiles in the same group, the impact coverage area of the missile is given by

$$\Omega = \bigcup_{i=1}^{n} \Omega_{mi} \tag{8}$$

To calculate (6) by numerical integration, it is necessary to define the boundary of the intercept coverage area. However, when there are more than one missile in the same group, the boundary of the covered area is difficult to be expressed. For this reason, this paper adopts the stochastic simulation theory to complete the calculation of interception probability through the following three steps.

Step 1: Produce a large number of real target points (y_t, z_t) according to the distribution density function (5).

Step 2: Decide whether the real target point is in the intercept coverage area and record it.

Step 3: Count all points in the intercept coverage area and divide that number by the total number of points in **step 1** to get the intercept probability.

In the stochastic simulation theory, the selection of the total number of random points has a great influence on the accuracy of calculation results. The Coefficient of variation ($C \cdot V$) of interception probability can reflect the calculation accuracy of the above algorithm. $C \cdot V$ is a normalized measure to evaluate the degree of dispersion of a statistic and is defined as the ratio of standard deviation to mean value.

The algorithm is run 1000 times, each time the algorithm produces N points and the interception probability is calculated.

The coefficient of variation curve as shown in Fig. 3.

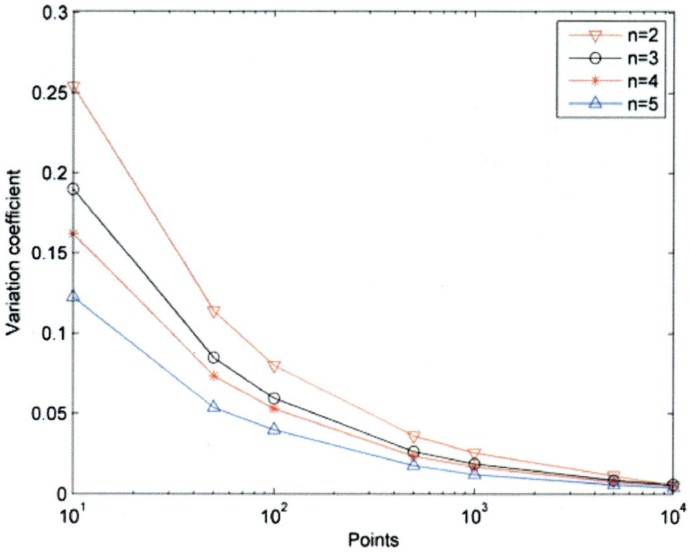

Fig. 3. The coefficient of variation curve

Figure 3 shows $C \cdot V$ of interception probability corresponding to different N. As can be seen from Fig. 3, when $n = 2, 3, 4, 5$, the coefficient of variation decreased with the increase of the total number of random points, which indicated that that the reliability improves when the total number of random points increases. However, the increase of the total number of random points will also lead to a rapid increase in the simulation time. Therefore, the total number of random points is set as 1000 in this paper.

3 Missile Location Optimization Based on IGWO

Obviously, when the number of missiles in the same group is only one, the intersection position of missiles in the center of the intersection plane will maximize the probability of interception. When the number of missiles in the same group is more than one, the optimal intersection position of the group of missiles cannot be obtained intuitively, and it needs to be calculated by the optimization algorithm. In addition, in this case, the interception probability can be regarded as a function whose independent variable is the position of the interchange point of the same group of missiles. To solve the optimal value of the function, at least $2n_i(i = 1, 2 \cdots, m)$ functions need to be considered, which requires high optimization ability of the optimization algorithm. Therefore, this paper uses an improved grey wolf optimization algorithm to solve the optimal interception probability.

Grey Wolf Optimizer (GWO) is a new group search method which simulates the social hierarchy mechanism and predation behavior of grey wolf in nature [5]. GWO algorithm is widely concerned because of its advantages of fast convergence and few parameters [6, 7].

However, the basic GWO algorithm, like other group search methods, is easy to fall into local optimality. In order to avoid falling into local optimum, Zhu proposed a hybrid GWO algorithm for function optimization based on the DE algorithm [8]. By introducing the dynamic evolutionary population technique into GWO algorithm, the global optimization ability of GWO algorithm is improved [9].

In this paper, the initial population is generated uniformly by Hammersley sequence, and then a nonlinear convergence factor adjustment strategy is proposed to enhance the global optimization ability of GWO algorithm.

3.1 The Traditional Grey Wolf Optimizer

The grading mechanism of gray wolf population is shown in Fig. 4, which consists of four grades: $\alpha, \beta, \delta, \omega$. Lower-ranking wolves strictly obey higher-ranking wolves. In the process of hunting, the three wolves with the best ability in turn are α, β, δ, these three wolves guide the other wolves to the target search.

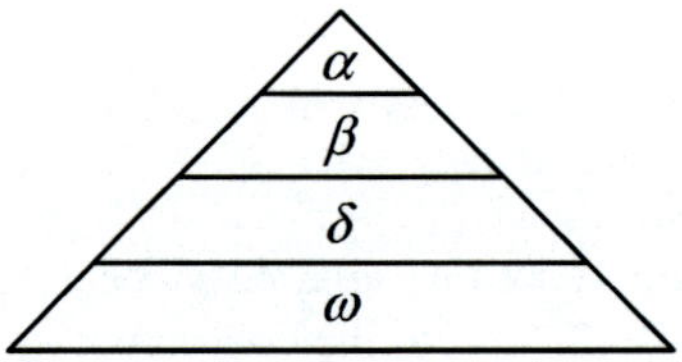

Fig. 4. The Wolf group level distribution

GWO defines a form of operation '.∗' which means to multiply the corresponding elements of two vectors.

The main algorithm to update the positions in GWO are (9) and (10).

$$\mathbf{D} = |\mathbf{C}.*\mathbf{X_P}(\mathrm{t}) - \mathbf{X}(\mathrm{t})| \tag{9}$$

$$\mathbf{X}(t+1) = \mathbf{X_P}(t) - \mathbf{A}.*\mathbf{D} \tag{10}$$

where $\mathbf{X_P}$ is the location of the prey, $\mathbf{X}(t)$ represents the position vector of the individual gray wolf in the t generation,**A** and **C** are coefficients, **A** and **C** can calculated by (11) and (12).

$$\mathbf{A} = 2\mathbf{a}.*\mathbf{r_1} - \mathbf{a} \tag{11}$$

$$\mathbf{C} = 2\mathbf{r_2} \tag{12}$$

where the elements of $\mathbf{r_1}$ and $\mathbf{r_2}$ are pseudorandom numbers from 0 to 1, the elements of $\mathbf{a}$ decrease linearly with the number of iterations from 2 to 0.

When $|\mathbf{A}| > 1$, the wolf pack spreads out, expanding the search area, corresponding to the global search, when $|\mathbf{A}| < 1$, the grey wolf pack shrinks the area surrounded, corresponding to local search.

The whole pack reaches the prey and attack by updating the position based on the best locations of the α, β, δ.

$$\mathbf{D}_\alpha = |\mathbf{C_1}. * \mathbf{X}_\alpha - \mathbf{X}(\mathrm{t})| \tag{13}$$

$$\mathbf{D}_\beta = \left|\mathbf{C_2}. * \mathbf{X}_\beta - \mathbf{X}(\mathrm{t})\right| \tag{14}$$

$$\mathbf{D}_\delta = |\mathbf{C_3}. * \mathbf{X}_\delta - \mathbf{X}(\mathrm{t})| \tag{15}$$

$$\mathbf{X_1} = \mathbf{X}_\alpha - \mathbf{A_1}. * \mathbf{D}_\alpha \tag{16}$$

$$\mathbf{X_2} = \mathbf{X}_\beta - \mathbf{A_2}. * \mathbf{D}_\beta \tag{17}$$

$$\mathbf{X_3} = \mathbf{X}_\delta - \mathbf{A_3}. * \mathbf{D}_\delta \tag{18}$$

$$\mathbf{X}(t+1) = (\mathbf{X}_1 + \mathbf{X_2} + \mathbf{X_3})/3 \tag{19}$$

The convergence of the GWO algorithm can be proved by Markov process. Therefore, after a finite number of iterations, the optimal solution can be obtained with the GWO algorithm.

3.2 Improved Grey Wolf Optimizer

3.2.1 Initialize the Grey Wolf Population with a Hammersley Sequence

For the swarm intelligence optimization algorithm based on population iteration, the quality of the initial population affects the global search speed and the quality of the algorithm greatly, and good diversity of the initial population can effectively improve the optimization performance of the algorithm [10]. However, the standard GWO algorithm uses random population initialization before iteration, which makes it difficult to ensure initial population diversity. Therefore, this paper proposes a population initialization method based on Hammersley sequence.

Hammersley sequence is a low-difference sequence that covers the whole space more completely, and its mathematical description is as follows [11]:

For any non-negative integer i and prime base p, Define a set of sequences

$$i = \sum_{k=0}^{r} a_k p^k \tag{20}$$

$$i = \sum_{k=0}^{r} a_k p^k \tag{21}$$

where $a_k \in [0, p-1]$.

$\Phi_p(i)$ can be determined by (20) and (21), and $\Phi_p(i) \in (0, 1)$.

In D-dimensional sampling space, take a sequence of isomers $p_1, p_2, \ldots, p_{D-1}$. Correspondingly, we can get the sequence $\Phi_{P_1}, \Phi_{P_2}, \ldots, \Phi_{P_{D-1}}$. The Hammersley point is defined as

$$x_i = \left(\frac{i}{N}, \Phi_{P_1}(i), \cdots, \Phi_{P_{D-1}}(i)\right), i = 0, 1, 2, \cdots, n-1 \tag{22}$$

Pseudo-random sequence (PN sequence) and Hammersley sequence (HM sequence) were used to generate 200 individuals of the initial population, respectively. As shown in Fig. 5, it is obvious that Hammersley sequences are more uniform than pseudo-random sequences. Therefore, using Hammersley sequence to generate the initial Wolf pack can uniformly cover the whole space, effectively avoiding the algorithm falling into the local optimal solution, and improving the global convergence.

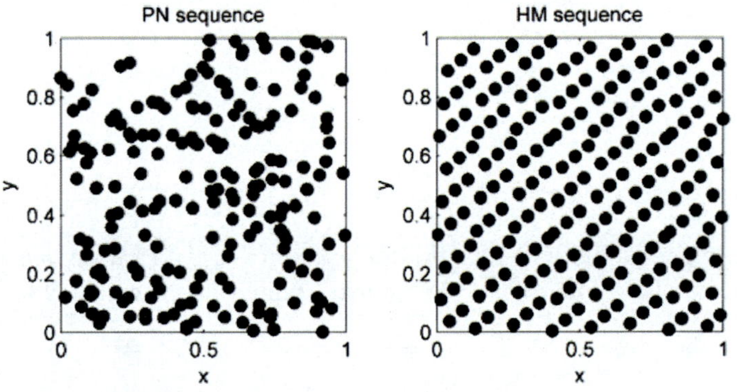

Fig. 5. Initial grey wolf population

3.2.2 Convergence Factor Adjustment Strategy

All swarm intelligence optimization algorithms have two kinds of behaviors in the optimization process: global search and local search. We can only quickly and accurately find the global optimal solution by balancing these two behaviors. Strong global search ability can maintain the diversity of the population, but increase the convergence time of the algorithm. Strong local search ability can realize small area accurate search, but can be easy to fall into local optimal. Therefore, it is necessary to coordinate the ability of global search and local search to increase the optimization performance of GWO algorithm.

In GWO algorithm, the value of **A** affects the global search and local search ability of GWO algorithm. As can be seen from (11), the change of **A** mainly comes from the change of the convergence factor **a**, which decreases linearly from 2 to 0 with the number of iterations. However, GWO changes non-linearly in the search process, and the linear decreasing mode of convergence factor **a cannot fully reflect the actual search** process [12]. Therefore, this paper proposes a new formula for the convergence factor of nonlinear change:

$$a(t) = 2\cos^{k}\left(\frac{\pi t}{2T}\right) \tag{23}$$

where a represents the element of $\mathbf{a}$,t is the number of current iterations, T is the maximum number of iterations, k **is the nonlinear regulation coefficient. In this paper, the** value of k is 0.1.

From (23), it can be seen that the convergence factor a changes non-linearly and dynamically with the increase of iteration times, effectively balancing the global search ability and local search ability.

3.3 Missile Location Optimization Process

The optimization process of the missile position on the intersection plane is shown in Fig. 6.

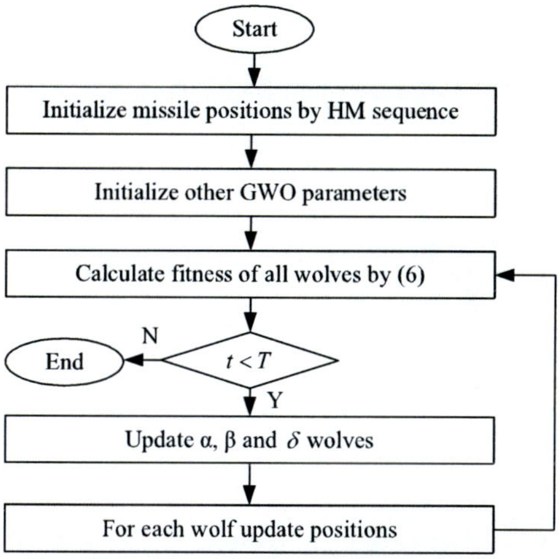

Fig. 6. Missile location optimization process

4 Simulation

4.1 Simulation of IGWO

In order to illustrate the superiority of IGWO algorithm, this paper compares IGWO algorithm with basic GWO algorithm and PSO algorithm, conducts optimization tests on common benchmark test functions, and conducts comparative analysis from the aspects of algorithm mean value and standard deviation. Table 1 gives the specific information of the benchmark functions, where F_1, F_2 are single-peak benchmark functions, F_3 is multi-peak benchmark function, and F_4 is fixed-dimension multi-peak benchmark function.

In order to reduce the impact of randomness on the results, each algorithm is used to solve the same function for 200 times. The average value of the 200 running results is used to evaluate the optimization ability, and the standard deviation is used to evaluate the stability of the optimization result. The optimization results of IGWO algorithm, GWO algorithm and PSO algorithm for the test function are shown in Table 2.

Table 1. Benchmark function details

Name	Test function	Dimension	Range	Min
Sphere	$F_1(x) = \sum_{i=1}^{n} x_i^2$	30	[−100,100]	0
Quartic	$F_2(x) = \sum_{i=1}^{n} ix_i^4 + random[0, 1)$	30	[−1.28,1.28]	0
Rastrigin	$F_3(x) = \sum_{i=1}^{n} \left[x_i^2 - 10\cos(2\pi x_i) + 10\right]$	30	[−5.12,5.12]	0
Goldstein Price	$F_4(x) = \left[1 + (x_1 + x_2 + 1)^2\left(19 - 14x_1 + 3x_1^2 - 14x_2 + 6x_1x_2 + 3x_2^2\right)\right] \times \left[30 + (2x_1 - 3x_2)^2 \times \left(18 - 32x_1 + 12x_1^2 + 48x_2 - 36x_1x_2 + 27x_2^2\right)\right]$	2	[−2,2]	3

Table 2. Simulation Result

Function	IGWO		GWO		PSO	
	Mean	Standard deviation	Mean	Standard deviation	Mean	Standard deviation
F_1	**8.817e-53**	**2.142e-52**	7.829e-31	1.375e-30	0.000134	1.55e-4
F_2	**3.4394e-04**	**1.4314e-04**	0.0015	7.6286e-04	0.0845	0.04
F_3	**0.04**	**0.8**	2.64	3.84	89	20
F_4	3.00002	2.262e-05	3.00004	4.599e-05	**3**	**0**

As can be seen from Table 2, compared with GWO algorithm, IGWO algorithm is superior to GWO algorithm in both mean value and standard deviation. Compared with PSO algorithm, except F_4, the global optimization ability and optimization stability of IGWO algorithm are obviously better than PSO algorithm. Therefore, we believe that the IGWO algorithm is effective, showing obvious advantages in finding the global optimal solution and algorithm stability.

Figure 7 shows the convergence curves of three algorithms for different benchmark functions. As can be seen from Fig. 7, compared with GWO and PSO algorithm, IGWO has faster convergence speed and higher convergence accuracy.

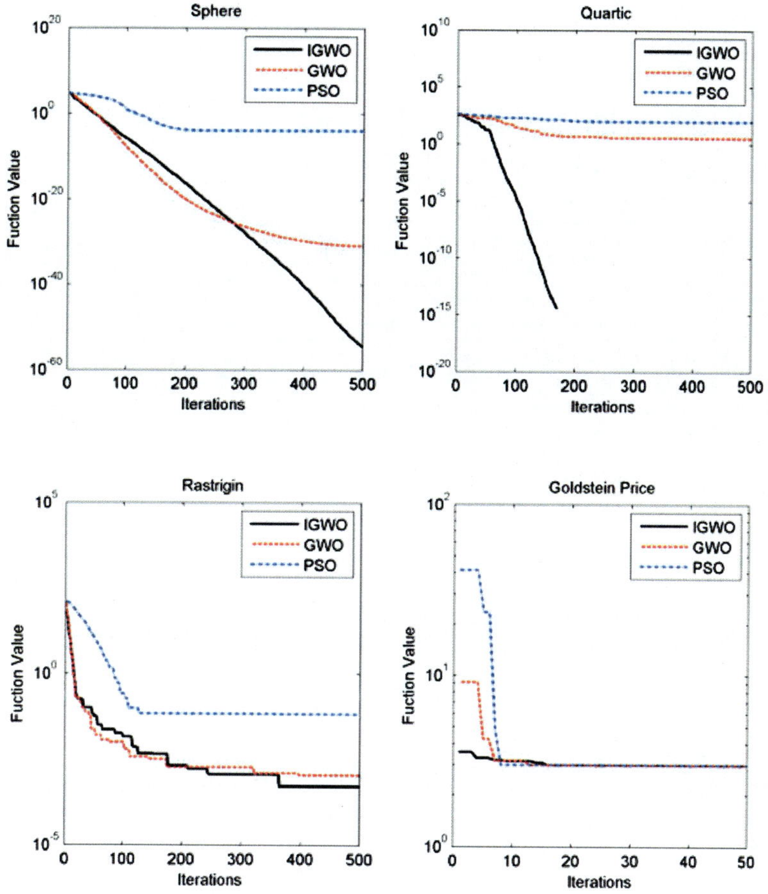

Fig. 7. Function convergence curve

4.2 Simulation of Multi-projective Coverage Interception Strategy

When the indication error exists, the main factors that affect the interception accuracy are guidance deviation, missile killing range, and the number and location distribution of missiles. This paper mainly studies the influence of the number and location distribution of missiles on the interception probability when the guidance deviation and the killing range of missiles are determined.

Therefore, parameters related to guidance deviation and missile killing range are set as follows

$$\left[\sigma_y, \sigma_z, r_{yz}\right] = \left[\sqrt{2}\mathrm{m}, 1\boldsymbol{m}, 1/\sqrt{2}\right] \tag{24}$$

$$R = 1\text{m} \tag{25}$$

Figures 8 and 9 show the optimal coverage with 3 and 4 missiles in the same group. In the case of limited number of missiles, by adjusting the position of the same group of missiles on the interception plane, the possible actual target points are covered to the maximum extent. This indicates that the interception strategy proposed in this paper is effective.

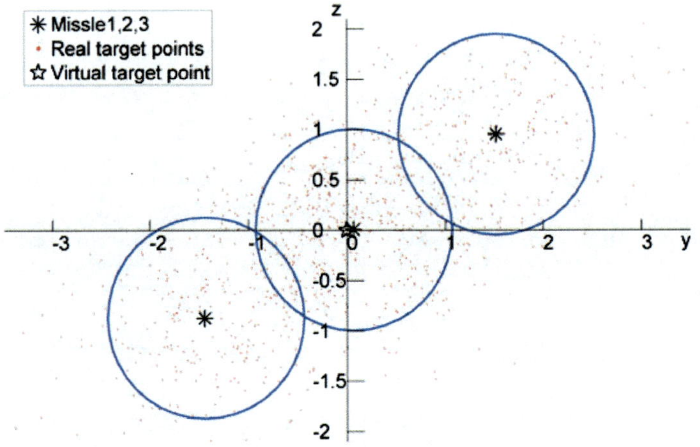

Fig. 8. Beat coverage with 3 missiles in the same group

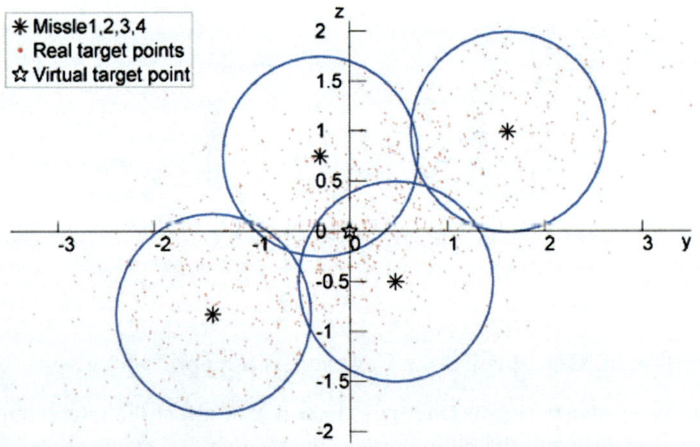

Fig. 9. Beat coverage with 4 missiles in the same group

Table 3 shows the optimal interception probability corresponding to different coverage interception strategies when the number of missiles is fixed. It can be seen that for a given number of missiles, the interception probability can be effectively improved by changing the coverage interception strategy, adjusting the number of missiles in different groups and optimizing the missile position. In addition, firing only one missile at a time in each group or using all missiles in the same group does not yield optimal interception probabilities. Therefore, it is also necessary to determine the number of missiles in different batches.

Table 3. Optimal interception probability

Total number of missiles	Missile group	Interception probability (%)
$n = 1$	$n = n_1 = 1$	**35.9**
$n = 2$	$n = n_1 + n_2 = 1 + 1$	58.9
	$n = n_1 = 2$	**61.2**
$n = 3$	$n = n_1 + n_2 + n_3 = 1 + 1 + 1$	73.7
	$n = n_1 + n_2 = 1 + 2$	**75.1**
	$n = n_1 = 3$	74.1
$n = 4$	$n = n_1 + n_2 + n_3 + n_4 = 1 + 1 + 1 + 1$	83.12
	$n = n_1 + n_2 + n_3 = 1 + 1 + 2$	84.06
	$n = n_1 + n_2 = 1 + 3$	83.4
	$n = n_1 + n_2 = 2 + 2$	**84.95**
	$n = n_1 = 4$	83.12

5 Conclusion

The interception probability model of multiple missiles in the same group is established according to the stochastic simulation theory, and the influence of the total number of random points on the calculation accuracy of the interception probability is analyzed.

By using Hammersley sequence to generate the initial population evenly, the search space was covered effectively. This paper proposes a strategy to adjust the convergence factor of GWO. By controlling the change speed of convergence factor, the global search ability and local search ability are balanced effectively. From the optimization simulation results of several benchmark test functions, it can be seen that the search accuracy and convergence speed of the improved algorithm are greatly improved.

Indicating error has a great influence on interception accuracy in laser beam guidance system. In order to solve this problem, this paper uses the improved gray Wolf algorithm to optimize the intersection position of multiple missiles in the same group, which effectively improves the interception probability of the group of missiles. When the total number of missiles is determined, the number of different batches of missiles is adjusted to further improve the overall interception accuracy.

References

1. Guodong Q, Jianqiao Y, Xiaolin A (2019) Multi-missile coordination high precision guidance and control method for beam-riding guidance. J Phys: Conf Ser
2. Zaikang Q, Qunli X (2005) Guided weapon control systems. Beijing (Chinese)
3. Tengfei Z, Hexin Z, Junjun H (2015) A review on development and application of laser-guided weapons. Electron Opt Control 66–71+98 (Chinese)
4. Shouchun L (2013) Analysis of surface-to-air missile weapon system. Beijing (Chinese)
5. Mirjalili S, Mirjalili SM, Lewis A (2014) Grey wolf optimizer. Adv Eng Softw 69:46–61
6. Song HM, Sulaiman M, Mohamed M (2014) An application of grey wolf optimizer for solving conbined economic emission dispatch problems. Int Rev Model Simuls 7(5):838–844
7. El-Gaafary A, Mohamed Y, Hemeida A et al (2015) Grey wolf optimization for multi input multi output system. Univers J Commun Nerworks 3(1):1–6
8. Aijun Z, Chuanpei ZL et al (2015) Hybridzing grey wolf optimization with differential evolution for global optimization and test scheduling for 3D stacked SoC. J Syst Eng Electron 26(2):317–328
9. Saremi S, Mirjalili SZ, Mirhalili SM (2015) Evolutions population dynamics and grey wolf optimazer. Neural Comput Appl 26(5):1257–1263
10. Haupt R, Haupt S (2004) Practical genetic algorithm. Wiley, New York
11. Niederreiter H, Yeo AS (2013) Halton-type sequences from global function fields. Sci China Math 1467–1476
12. Pan, TS et al (2015) A communication strategy for paralleling grey wolf optimizer. In: International conference on genetic & evolutionary computing Springer International Publishing

Design and Optimization of the Flexible Support Structure for Space Mirror

Yixuan Sun(✉), Shikui Luo, Jie Bai, Zijia Liu, and Shaofan Tang

Beijing Institute of Space Mechanics and Electricity, Beijing 100094, China
sundae90@126.com

Abstract. The mirror subassembly is the key component of the space optic remote sensor and the mirror shape is always required to reach high accuracy due to the crucial importance to the system's imaging quality. However, the mirror shape is usually influenced by gravity, temperature, forced displacement and launch vibrations to distort after leaving earth. The support structure of the subassembly plays an important role to minimize the impact of above factors while the conflict of the mirror's static accuracy and the subassembly's dynamic strength needs to be studied. Therefore, based on a Φ 260 mm-aperture mirror subassembly, a whole set of flexible support structures was researched. Combining with the conventional ring-type lateral support structure, a discrete-type flexible mounting scheme was proposed. Materials of the mirror and cell, the micro-stress mounting strategy and parameters of the flexure hinges were established aiming at maintaining high static precision of the mirror under multi-conditions, and the resilient connector was optimized by stress deconcentration and viscoelastic materials damping to obtain adequate dynamic strength margin. Finally, the comprehensive performance of the subassembly was verified eligible by finite element analysis that the mirror surface error RMS was superior to 0.002 λ under objective static conditions with enough safety margin in the harsh vibrations. In conclusion, the flexible support structure brought forward by this article is feasible which has general compatibility for other small-size space mirror.

Keywords: Space mirror · Support structure · Flexible hinge

1 Introduction

The optical mirror is the core component of the space remote sensor whose surface figure accuracy is of critical importance for the system's image quality. However, the surface fidelity usually suffers from static precision aberrations caused by gravitational release, thermal deformations, and alignment errors during the deployment or working procedure [1].

Therefore, developing a reasonable support structure is the key approach to achieve adequate mirror surface fidelity under multiple complex conditions [2]. The support structures generally involved can be divided into 3 types: lateral support, central support and bottom support, of which the first way is the simple one that is often adopted for

H. Ding (ed.), *Aerospace Mechatronics and Control Technology*, Springer Aerospace Technology,
https://doi.org/10.1007/978-981-16-6640-7_10

mirrors of small aperture. Typically, the mirror is set in a metal ring-type substrate and rigidly mounted on the main frame. But when the mirror's size increases to 260 mm or greater, this strategy can hardly meet the requirements on the mirror's surface precision claimed by the optic system [3]. In order to reduce the surface error brought by temperature and assembly as well as ensure adequate strength to stick up to gravity effect and dynamic environment, research about different kinds of flexible mirror support structure were carried out successively [2–6].

In this article, a Φ260 mm-diameter space mirror was expected to maintain surface figure error root mean square (RMS) at the high precision of less than λ/60 and survive in launch vibrations at the same time, thus the idea of combing lateral support with flexible mounting structure was proposed and a vibration mitigation method by application of the viscoelastic damping adhesive was studied. Finally, the whole approaches were verified by the finite element method to be effective and satisfactory.

2 Conventional Lateral Support Structure

A typical method to laterally support the small mirror was using distributed adhesives to connect the mirror with an annular cell at the lateral side, see Fig. 1. The mirror subassembly consisted of mirror, annular cell, axial clamps, radial glue spots and axial glue spots. The mirror was upholded by the cell through the radial glue while the axial glue only worked to limit the moving range for protection [7].

Usually, the cell was made of titanium alloy and the clamps were connected to the main frame rigidly. Under this circumstance, when temperature goes up or down, the mirror will be strongly dragged by the cell to expand or contract as the coefficient of thermal expansion (CTE) of the two was rather different, and when the subassembly is mounted with inevitable alignment error or the fitting surface deforms, the mirror will also be strongly forced to distort. All these factors may cause surface figure degradation of the mirror.

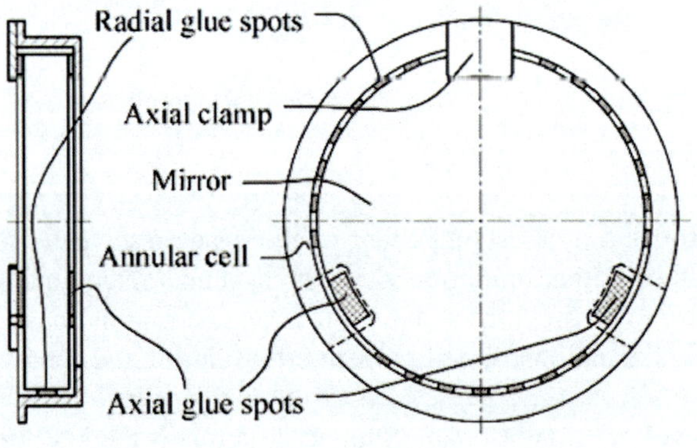

Fig. 1. Typical mirror support structure.

3 Study of the Flexible Support Structure

3.1 Research Goal

The mirror discussed in this paper is 260 mm in diameter which was required to keep the surface figure RMS superior to $1/60\lambda(\lambda = 632.8$ nm) throughout the whole lifecycle and withstand the launch vibrations without plastic deformation or failure.

The working conditions taken into consideration include: (1) 1G (9.8 m/s^2) gravity in the assembly direction; (2) Temperature rise of 5 °C; (3) 0.02 mm translational or 10" rotational aberration of the fitting surface; (4) Target level of random vibration.

With such high precision demand, the goal of this paper is to reduce the surface error RMS caused by (1) ~ (3) to less than 0.002λ and ensure the stress safety margin of all parts to be higher than (3 in 4).

3.2 Kinematic Support Principle

To largely release the distortion stress passed from the main frame, the kinematic support method was investigated. As Fig. 2 presents, the mirror-cell-clamp (MCC) module was mounted on the main frame through three connectors. In order to set the module at a determinate position free of redundant restrictions, a statically determinate structure with zero degree of freedom (DOF) was proposed.

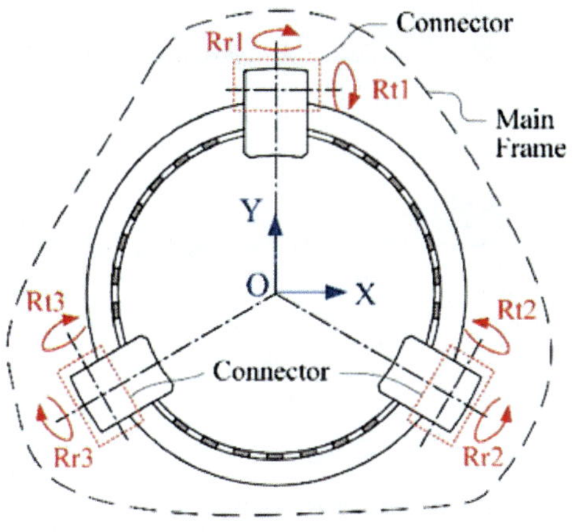

Fig. 2. 2 DOF of the MCC's connector.

The DOF of the spatial mechanism without general constraint is shown as follow [8]

$$W = 6(n - 1) - \sum_{\mathrm{k}=1}^{5} (6 - \mathrm{k})P_k \tag{1}$$

where W denotes the whole DOF of the mechanism, $(n - 1)$ denotes the number of moving parts, P_{k} denotes the number of k-level kinematic pair and k denotes the number of kinematic pair's constraints. As for the mirror subassembly in Fig. 2, each connector has a 2 DOF motion pair with the revolving DOF Rt and Rr which takes on 4 constraints. Therefore, $P_1 \sim P_3, P_5 = 0, P_4 = 3$, and $(n - 1) = 1$. It can be deduced that the MCC's DOF is zero which can meet the requirements for exact restriction.

3.3 Preliminary Design

3.3.1 Structure Composition

Due to the inevitable frictions, fit clearance, transmission error and creep issues of the rigid motion pairs would greatly increase the design complexity and impair the system accuracy [9], the specially designed resilient connector which could realize high-precision movement in a small range [10] will be used to simulate the required rotational DOF. The mirror subassembly consisted of the MCC module and three resilient connectors, which adopted 2 orthogonal circular-arc flexible hinges to simulate the 2 rotational DOF with relatively high rigidity and good machining accuracy, see Fig. 3.

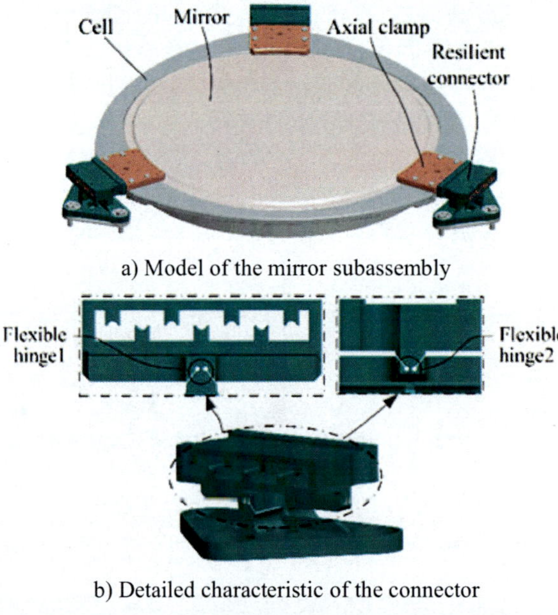

Fig. 3. Preliminary structure composition

3.3.2 Material Selection

The silicon carbide (SiC) ceramics has relatively high rigidity, low CTE, moderate density and good dimensional stability, so it was choosed as the mirror material to design a lightweight and stable optical element. To reduce the optical distortion caused by the mirror and cell's inconsistent expansion or contraction, materials of both need to be thermal adaptive. Carbon fiber reinforced silicon carbide ceramic matrix (C/SiC) composites, as a promising structural material in the aerospace field, has relatively high specific strength with CTE close to the SiC ceramics. So, it was choosed as the cell material to provide reliable and compliant support for the mirror.

As the other parts were not joint with the mirror directly and expected to be high-strength and lightweight, so the titanium alloy was choosed in synthetical consideration.

Materials of the main components of the mirror subassembly and their detailed performance parameters are listed in detail in Table 1. In a whole, the mirror subassembly only weighs 2.1 kg.

Table 1. Performance parameters of materials.

Material	Elastic modulus E/(Gpa)	Poisson's ratio μ	Density ρ/(g/cm^3)	CTE α/(μK^{-1})
SiC	350	0.17	3.12	2.5
C/SiC	70	0.3	2.5	1.6
Titanium alloy	110	0.32	4.5	9.1

3.3.3 Mounting Principles

Based on the flexible support structure, a micro-stress mounting strategy was proposed to further reduce the initial discomfort of the mirror brought by assembly. The overall idea is firstly, fit the MCC module to the main frame at the theoretical position through a special non-contact tooling, secondly locate the connectors by the corresponding clamp and assemble them to the main frame with bolts, then remove all the temporary toolings, finally inject the glue filler to the jagged clearance through the injection channels and holes with the help of the dedicated tooling and wait the glue filler became dry, see Fig. 4. That is, the clamps and connectors will join with tiny stress.

3.3.4 Parameter Design of the Flexible Hinge

The parameters of the flexible hinge are associated with the bracing stiffness of the connector which need to be suitable to resolve the conflict between error unloading capability and dynamic carrying capacity. The rotational stiffness of the circular-arc flexible hinge K is the key to design which can be expressed as [11]

$$K = M_Z/\alpha_Z \approx \left(2Ebt^{5/2}\right)/\left(9\pi R^{1/2}\right) \tag{2}$$

where M_Z denotes the torque exerted on the hinge, α_Z denotes the rotation angle of the hinge, E denotes the material's elastic modulus, b denotes the width of the hinge, t denotes the thickness of the thinnest place of hinge section, and R denotes the radius of the circular arc, see Fig. 5.

According to Eq. (2), the argument t impacts most which need to be focused on in subsequent optimization. Therefore, the value of b and R were pre-set based on

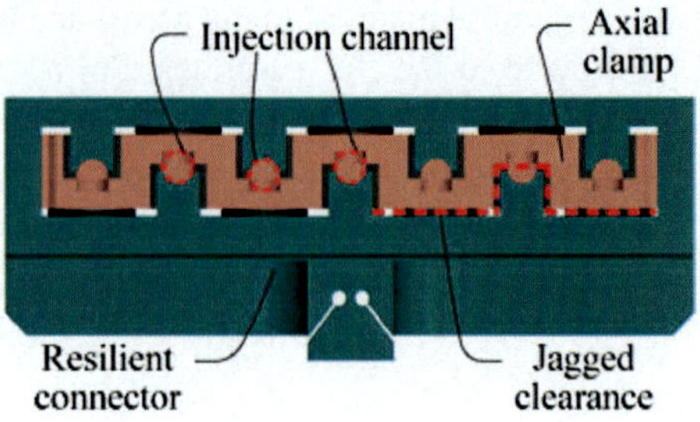

a) Make up of the clamp-conncetor module

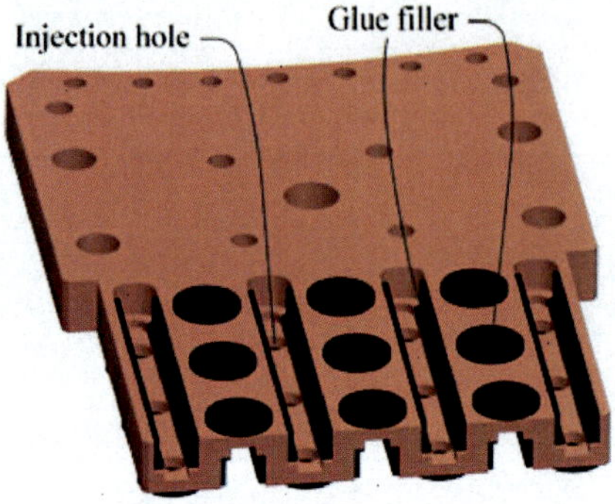

b) The reselient connector hidden

Fig. 4. Assembly elements of the clamp-connector module.

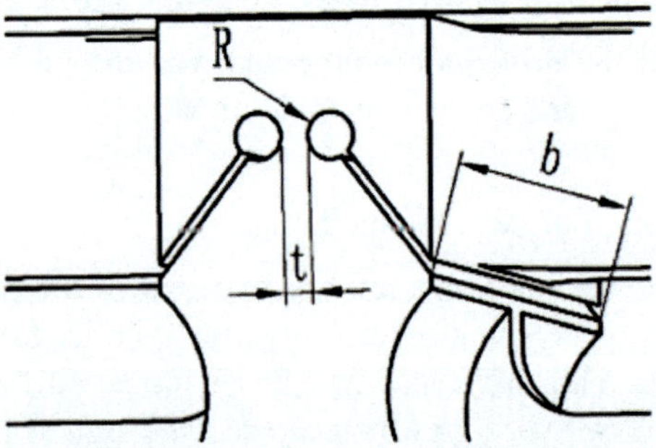

Fig. 5. Parameters of the circular-arc hinge.

the structure size and process requirements. Suppose b_i, R_i and t_i denotes the relative parameters of the flexible hinge i (see Fig. 3), $b_1 = 28$ mm, $b_2 = 48$ mm, and $R_1 = R_2 = 0.5$ mm. With the aid of finite element analysis (FEA), the optimal value of t_1 and t_2 were finally found at 0.6 mm to meet the forced displacement unloading need.

4 Analysis and Improvement

4.1 Static Precision Analysis

To show the beneficial effects of the support structure proposed by this paper, the surface error RMS of the following two types of mirror subassembly were analyzed by the FEA method respectively: (A) Typical type: the subassembly fitted with hard link, using the titanium alloy cell; (B) New type: the subassembly fitted with the special resilient connectors, using the C/SiC cell.

Then the following 5 working conditions were simulated (based on Fig. 6): (a) A ~ C fixed with 1G gravity along -Y; (b) A ~ C fixed with temperature rise of 5 °C; (c) B and C fixed while A was given 0.02 mm displacement + Z; (d) B and C fixed while A was given 10" displacement about + X; (e) B and C fixed while A was given 10" displacement about + Y.

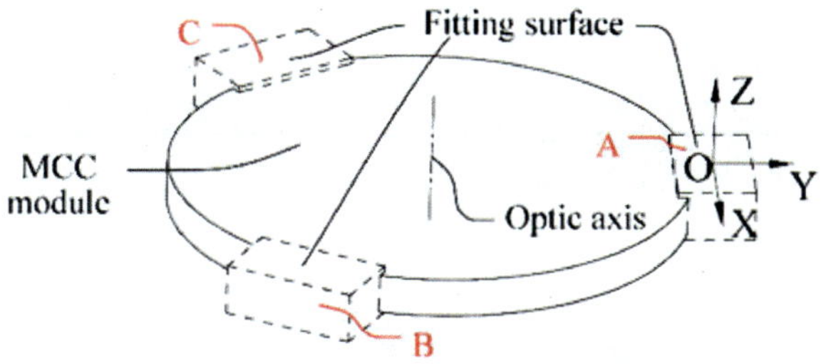

Fig. 6. Definition of the simulation.

By simulation, the mirror's surface figure RMS was obtained, see Table 2. Clearly that the typical type cannot fit the bill in the (b), (c) and (e) conditions with RMS error greater than the requested 0.002λ, while the new type structure behaved as expected whose results all satisfied the demand.

Table 2. Surface RMS of the 2 types in 5 working conditions.

RMS /λ	(a)	(b)	(c)	(d)	(e)
(A)	0.00059	0.00469	0.00690	0.00010	0.00304
(B)	0.00080	0.00155	0.00017	0.00002	0.00009

4.2 Dynamic Strength Analysis

The constraint modal of the subassembly was carried out based on FEA and the fundamental frequency was found at 118 Hz showing as the mirror's translation along the optical axis. Then, the frequency response in 3 directions from 20 to 2000 Hz was simulated. Overall, the Z-direction response at 118 Hz manifested as the harshest condition which has the highest stress level.

On this basis, the random response of acceleration and stress under given level of random vibration (5G RMS from 20 to 2000 Hz) was obtained. Overall, there was only one risk area of the subassembly (see Fig. 7) whose Von-Mises stress level was high to the yield limit. The maximum stress RMS appeared at the section center of the flexure hinge and reached 258 Mpa. According to the 3σ clipping rule [12], the maximum stress was 774 Mpa, close to the titanium alloy's yield limit 810 Mpa. Therefore, the safety margin is inadequate and need to be improved urgently.

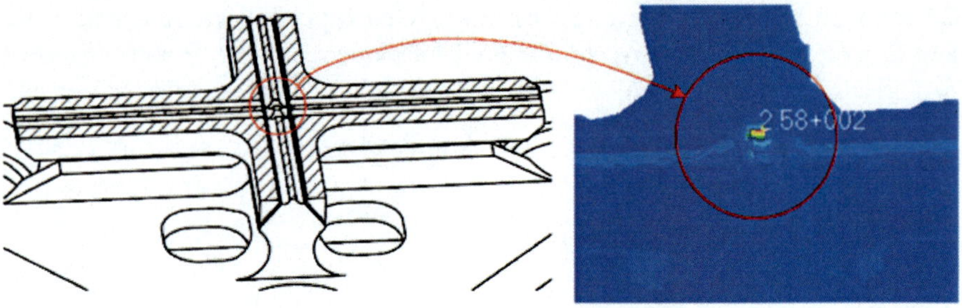

Fig. 7. Maximum stress area of the resilient connector.

4.3 Improvement

The stress concentration and the flexure design caused high stress level in the connector which could be solved following the two ideas: (a) removing the materials of the stress concentration region; (b) adopting the damping anti-vibration method.

In terms of the first approach, a hole cutting through the flexure hinge centre was set which removed the top 80% high stress solid, see (a) in Fig. 8. For the second approach, shear deformation occurred in the viscoelastic damping materials to turn the vibration mechanical energy into heat and dissipate, thus to reduce the response [13]. The viscoelastic materials' low elastic modulus and high shear loss factor in specific temperature and frequency are critical parameters for the damping effect [14]. In order to take full advantage of the shear properties of the damping material, the adhesive was arranged along the tangent of the flexure hinge to bear the shear force, see (b) in Fig. 8.

By calculation, the maximum stress of the resilient connector was reduced to 243Mpa (based on 3σ rule) with safety margin of 3.33 which verified the subassembly's dynamic strength to be eligible.

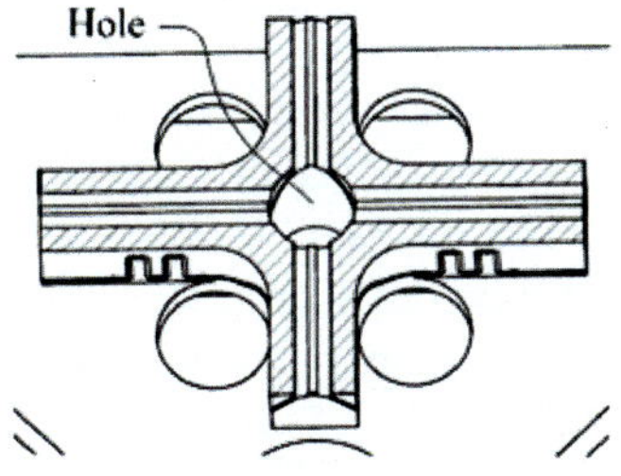

a) First approach

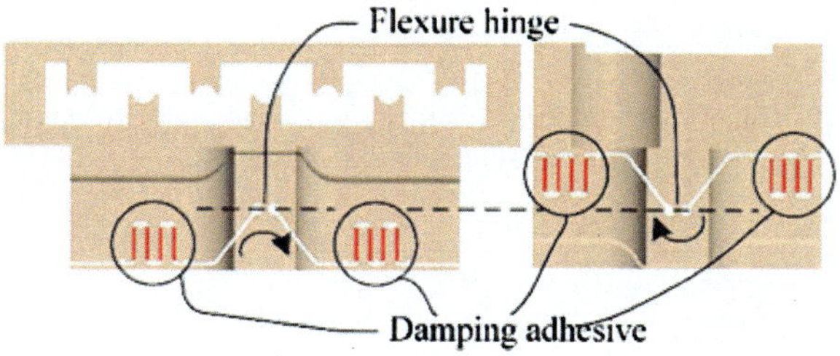

b) Second approach

Fig. 8. Improvement of the resilient connectors.

4.4 Recheck of the Static Precision

The improved structure's static precision related with gravity, temperature and assembly need to be re-assessed in the conditions same as Sect. 4.1. Table 3 compares the surface figure error RMS of the original and the improved structure. It can be verified that the improvement didn't cause bad impact on the static precision of the subassembly and the modified scheme was feasible.

Table 3. Surface RMS of the original and improved structure.

RMS /λ	Gravity	Temp	0.02 mm along + Z	10" about + X	10" about + Z
Original	0.00080	0.00155	0.00017	0.00002	0.00009
Improved	0.00125	0.00155	0.00003	0.00001	0.00013

5 Conclusion

This paper studied the flexible support structure for a Φ260mm-diameter space mirror. In order to suffice the comprehensive requirements of static precision and dynamic strength under multi-conditions, a statically determinate discrete support scheme was proposed based on the kinematic analysis method, combining the conventional ring-type lateral support structure with resilient connectors disposed at the periphery. Targeting at minishing the optical degradation against thermal variation and forced displacement from the fitting area, parameters of the connector's flexure hinge were optimized, and the micro-stress mounting strategy was applied while materials of the mirror and cell were creatively choosed to be CTE consistent. Later, the comprehensive performance related with gravity, temperature and assembly was evaluated to be eligible. As the safety margin in random vibration was insufficient, the resilient connector was optimized to be stress deconcentrated and damp-increased with viscoelastic materials, which notably improved the safety margin to be eligible with undisturbed error unloading capability. The flexible support structure put forward by this paper is of reference meaning for design of other the small-size space mirror subassembly.

Acknowledgements. Financial supports from the National Key Research and Development Program of China (No. 2016YFB0500501) are gratefully acknowledged.

References

1. Rausch P, Verpoort S, Wittrock U (2015) Unimorph deformable mirror for space telescopes: design and manufacturing. Opt Express 23(15):19469–19477
2. Wang ZS, Zhai Y, Mei G et al (2010) Design of flexible support structure of reflector in space remote sensor. Opt Precis Eng 18(8):1833–1841
3. Wang WP, Lv TY, Liu XY et al (2019) Flexure support design and optimization for middle-small spherical mirror. J Chang Univ Sci Technol 42(3):7–10
4. Liu M, Zhang LZ, Li X et al (2018) Design of flexure support of space compact reflector subassembly and dynamic analysis. Opto-Electron Eng 45(5):1–10
5. Wang ZS, He X, Fu LL (2011) Design and analysis for a minitype mirror supporting structure. Opt Tech 37(6):686–690
6. Li YP, Wang Z, Sha W et al (2018) Structural design of primary mirror subassembly for spatial gravitational wave telescope. Infrared Laser Eng 47(8):1–7
7. Luo SK, Cao DJ, Lan LY et al (2016) Support technique of space mirror based on diversity between apparent Young modulus and shear modulus. Spacecr Recover Remote Sens 37(1):41–47
8. Zhang YX, Chen DS, Liao ZF et al (2003) Exploration on the formula of degree of freedom of space mechanism. J Chongqing Univ 26(9):53–55
9. Shi RC (2013) Optimal design and system development of a flexible parallel high-precision pointing mechanism [D]. Harbin: Harbin Institute of Technology
10. Tseytlin Y (2012) Note: rotational compliance and instantaneous center of rotation in segmented and V-shaped notch hinges. Rev Sci Instrum 83(2):026102
11. Yang SY (2008) Analysis of geometrical parameters of circular arc flexible hinge influence on its rigidity. Coal Mine Mach 29(7):81–82
12. Zhu XW, Zhang SJ, Ning ZG et al (2014) Applications of 3σ clipping rule in the fatigue damage accumulation analysis. Spacecr Environ Eng 31(6):609–613
13. Baker M (2007) Analysis method to support design for damping. Eng Comput 23(1):1–10
14. Qiu JH, Fan B, Lian HD (2015) Research on the application of viscoelastic constrained layer damping in space large aperture mirror. Spacecr Recover Remote Sens 36(2):32–38

Experimental Research on the Influence of the Bulk Swirl on the Compressor

Feng Lu-ning(✉), Cheng Bang-qin, Wang Jia-le, Tao Ping, and Wu Qiang

Aeronautical Engineering College, Air Force Engineering University, Xi'an 710038, China
1450436728@qq.com

Abstract. In order to explore the effect of the overall vortex flow distortion on the performance of the low-speed axial compressor, this paper carried out the compressor characteristic experiment under the overall vortex flow distortion generator based on the overall vortex flow distortion generator, and measured the performance of the swirl flow distortion generator. The generated swirl field obtains the pressure ratio and efficiency characteristics of the compressor under the conditions of uniform intake and swirl intake. The results show that the overall vortex flow distortion increases the pressure ratio of the compressor, reduces the efficiency of the compressor, and at the same time causes the stability margin of the compressor to decrease.

Keyword: Bulk swirl · Low-speed axial compressor · Swirl distortion generator

1 Preface

In recent years, with the development of modern warfare, in order to meet the requirements of stealth and high mobility [1, 2], a series of complex inlets such as S-curved inlets have been applied to more and more fighters. While improving the performance of the fighter, the changes in the intake duct structure and the complexity of the fuselage made the swirl distortion generated by the engine inlet more serious, which adversely affected the stable operation of the engine.

Swirl distortion is mainly divided into four types [3]: overall vortex flow, counter vortex flow, concentrated vortex flow (winding vortex flow) and transverse vortex flow, of which the overall vortex flow occurs most frequently. And it has a greater impact on engine performance and stability.

In the past few decades, a lot of research work has been carried out at home and abroad to explore the effects of swirl distortion. Sheoran et al. [4–6] designed a vortex flow distortion enerator, and carried out numerical and experimental research on the effect of vortex flow on compressor performance. Domestic Jiang Jian et al. [7–9] designed a vane-type swirling distortion generator to generate counter-vortex flow. Tu Baofeng et al. [10–12] designed a vane-type swirling distortion generator and carried out numerical simulation and experimental research. Cheng Bangqin et al. [13–18] designed a chamber-type swirl distortion generator and swirl distortion net, and carried out corresponding numerical simulations.

H. Ding (ed.), *Aerospace Mechatronics and Control Technology*, Springer Aerospace Technology,
https://doi.org/10.1007/978-981-16-6640-7_11

Based on the overall swirl flow distortion generator designed and processed, this paper carried out an experimental study on the effect of overall swirl flow distortion on the performance of low-speed compressors. The downstream flow field of the swirl distortion generator was measured in detail, and uniform intake air was obtained. And compressor characteristics under swirling air intake conditions.

2 Laboratory Equipment

2.1 Low-Speed Axial Compressor Test Bench Structure

Figure 1 is a schematic diagram of the structure of the compressor laboratory, which is mainly composed of a bell mouth, a swirling distortion section, a stator section, a throttle valve, a stepping motor, a synchronous motor, a torque meter and a fluid pipeline. The power source of the compressor is a 30 KW AC motor, which is controlled by a frequency converter, and can obtain any speed from 0 to 3000 rpm. The inlet of the compressor adopts a bell mouth design to make the air intake uniform. After the air flow passes through the swirl distortion generator, it flows through the fairing, is compressed by the rotor and enters the exhaust section, and finally flows out of the compressor. A throttle valve controlled by a stepping motor is installed in the exhaust section to control the size of the intake air flow. The isolated rotor adopts an equal diameter design with a hub radius of 200 mm and a casing inner wall radius of 300 mm. There are 45 blades in total, a design speed of 3000 rpm, and a design pressure ratio of 1.025.

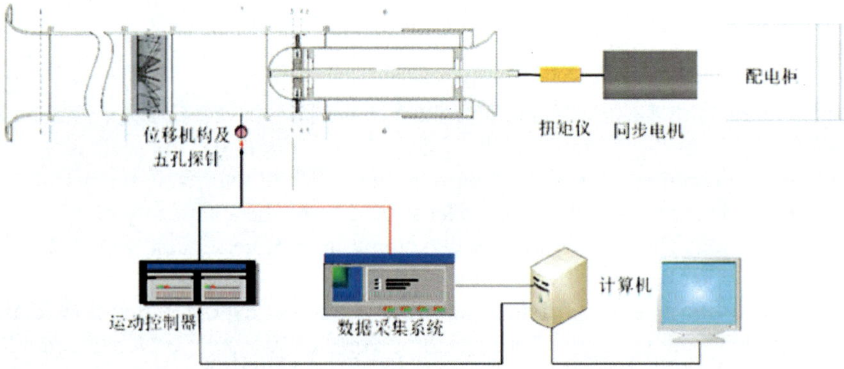

Fig. 1. Compressor experimental test system

2.2 Swirl Distortion Generator

The swirling distortion generator adopts a blade-type swirling distortion generator previously designed by this research group, and its specific design ideas can refer to the literature [17]. Figures 2 and 3 respectively show the three-dimensional effect diagram of the vortex flow distortion generator and the installation effect diagram of the test piece. The resin material is used for 3D printing, and post-treatment is carried out to increase the surface smoothness and anti-oxidation treatment.

Fig. 2. Three-dimensional diagram of the overall vortex flow distortion generator

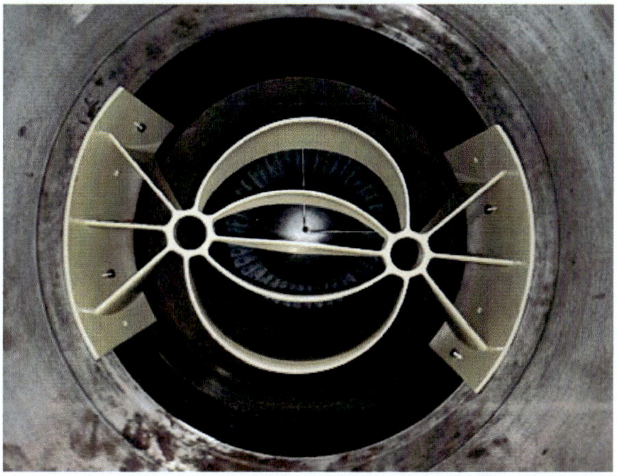

Fig. 3. Test piece for vortex flow distortion generator

2.3 Swirl Field Measurement Program

In the experiment, the flow field at a position 500 mm downstream of the swirling distortion generator was measured. Five-hole probes were arranged above and on the side of the pipe, and the five-hole probes were fixed on the displacement mechanism by clamps. As shown in Fig. 4, the experiment measures 12 circumferential positions, 10 radial positions, and a total of 120 measuring points. In the experiment, the rotation distortion generator is used to measure different circumferential positions, and the radial position of the probe is changed by controlling the displacement mechanism.

The swirl distortion evaluation index proposed by SAE is introduced here. The swirl distortion evaluation standard proposed by SAE is based on the swirl angle and uses three indexes and two sub-parameters to evaluate swirl distortion. The three indicators are the swirling sector, swirling intensity and swirling direction, and the two sub-parameters are the circumferential range and swirling sector.

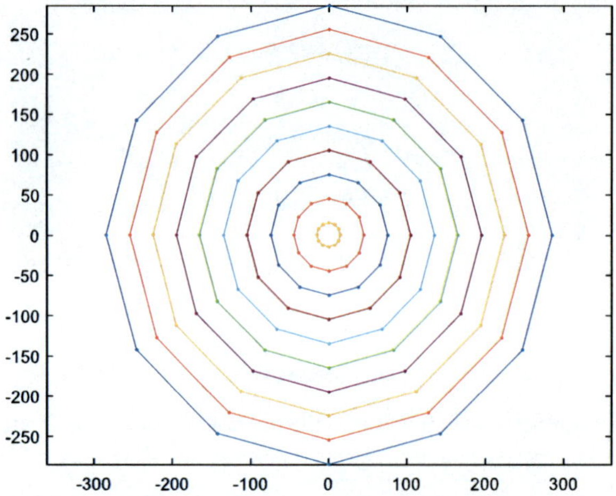

Fig. 4. Schematic diagram of measuring points

The definition of swirl angle is:

$$\alpha = \tan^{-1}\left(\frac{U_\theta}{U_x}\right) \tag{1}$$

U_θ Circumferential velocity component at a point on AIP
U_x The axial velocity component of a point on the AIP

3 Experimental Result

3.1 Swirl Field Characteristics

Ten measuring loops are set up in the flow field (Fig. 5). Figure 6 shows the swirl angle cloud diagram of the AIP after the air flows through the overall vortex flow distortion generator. Figure 7 shows the simulation results and the comparison of the experimental results. It can be seen from Fig. 7 that the simulation results are relatively close to the experimental results.

Figure 8 shows the changes and differences of swirl angles on different measuring rings. Here, we only take the swirl angle distributions on the first, fourth and seventh measuring rings as an example.

3.2 The Effect of Integral Vortex on Rotor Performance

The experiment measured three speed lines: 80%, 70% and 60% of the design speed. Under the action of the overall vortex flow, the change of the rotor characteristic line is shown in Figs. 9, 10 and 11. It can be seen that under the three speed lines, the pressure ratio characteristic line of the rotor moves to the upper right under the action of the reverse swirl flow, and the supercharging capacity increases slightly, but the reverse

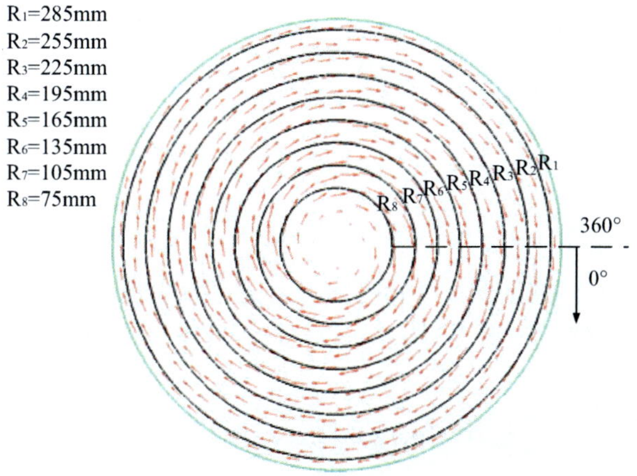

Fig. 5. Schematic diagram of AIP measuring loop

Fig. 6. Swirl angle cloud diagram at AIP after the overall vortex flow distortion generator

overall swirl flow causes the rotor stall point to advance and stable operation The margin decreases, and as the speed increases, the greater the loss of stability margin. At the three speeds, the stall point flow coefficient increased by 7.27%, 12.08% and 11.79% respectively. Affected by the reverse swirling flow, the efficiency characteristic line of the rotor moves downward. Under the three speed lines, the peak efficiency is reduced by 4.9%, 9.3% and 5.7% respectively. In addition, near the blockage point, compared with uniform intake, the efficiency characteristic line changes from steep to gentle under the action of reverse swirl.

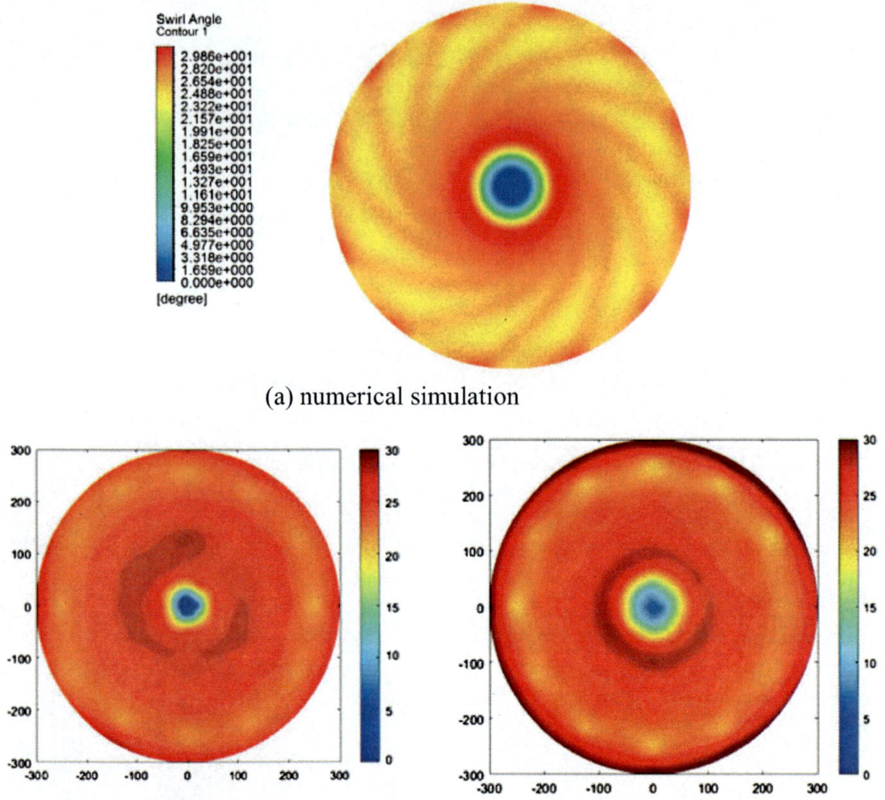

(a) numerical simulation

(b) Probe X to measure flow field

(c) Probe Z measures the flow field

Fig. 7. The Institute of Numerical Simulation and Experiments obtained the overall vortex flow angle cloud map

4 Conclusion

Based on the overall vortex flow distortion generator, this paper carried out an experimental study on the effect of the overall vortex flow on the performance and stability of the rotor. The specific conclusions are as follows:

(1) The designed overall vortex flow distortion generator can accurately simulate the overall vortex flow, and the error between experimental data and numerical simulation data is small;
(2) The overall vortex flow improves the pressurization capacity of the compressor, but reduces the stable working margin of the rotor, which advances the stall point of the rotor. At 60%, 70% and 80% of the design speed, the stall point flow increases respectively. 7.27%, 12.08% and 11.79%.

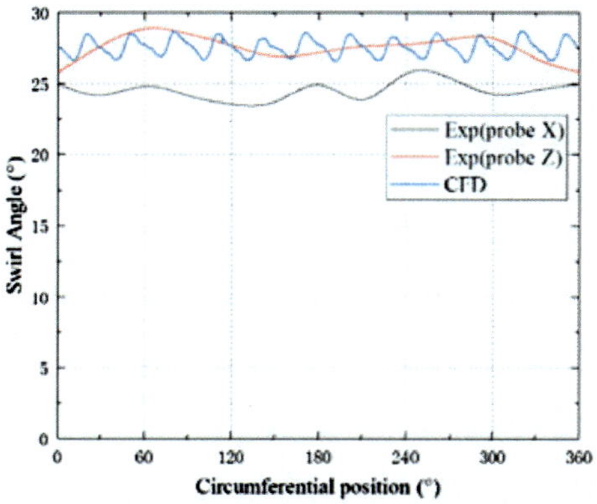

(a) R1

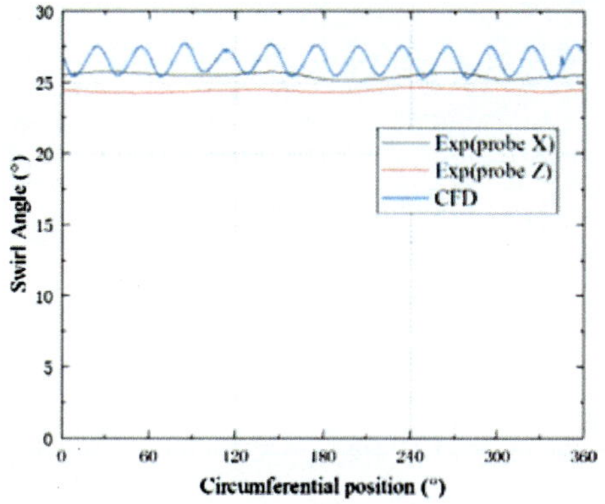

(b) R4

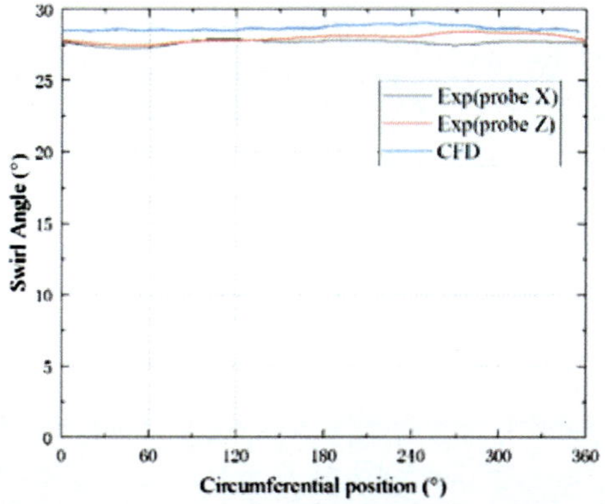

(c) R7

Fig. 8. AIP and swirl angle distribution on the measuring ring under the action of the overall vortex flow

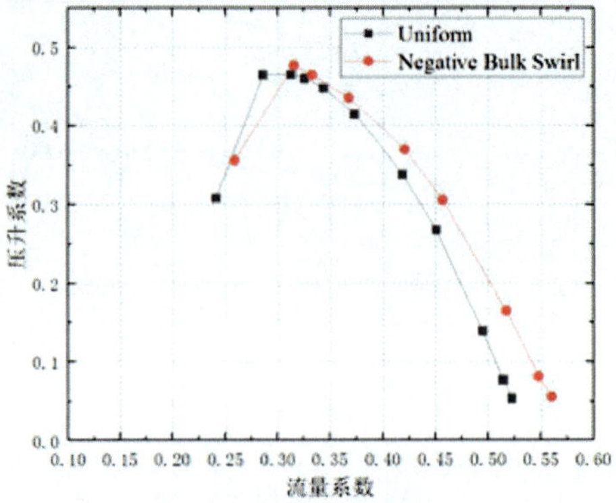

(a) Pressure ratio

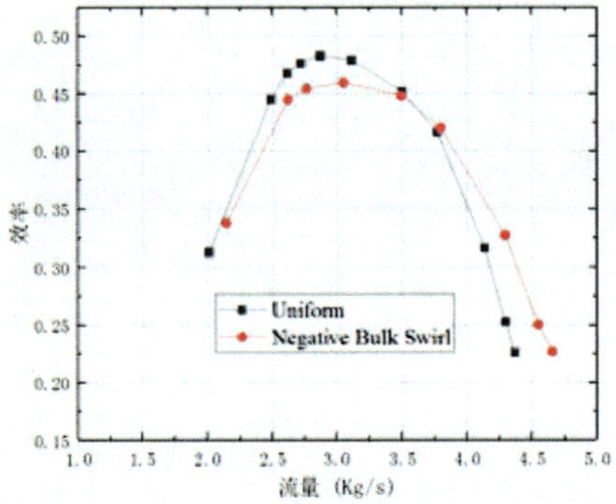

(b) Efficiency characteristics

Fig. 9. The effect of the overall swirl flow on the rotor characteristics (1800 rpm)

(3) The overall vortex flow has a negative impact on the rotor efficiency. At 60%, 70%, and 80% design speeds, the peak efficiency is reduced by 4.9%, 9.3%, and 5.7%, respectively.

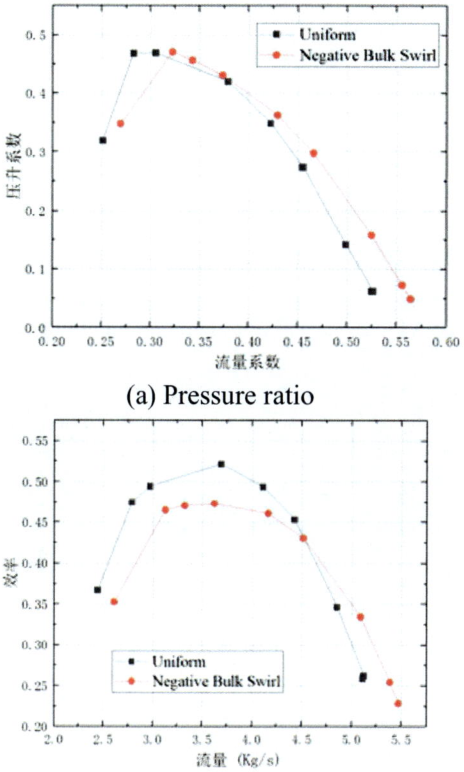

(a) Pressure ratio

(b) Efficiency characteristics

Fig. 10. The effect of the overall swirl flow on the rotor characteristics (2100 rpm)

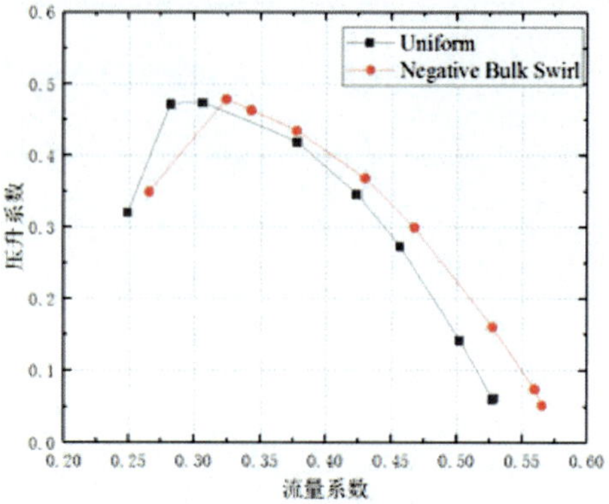

(a) Pressure ratio

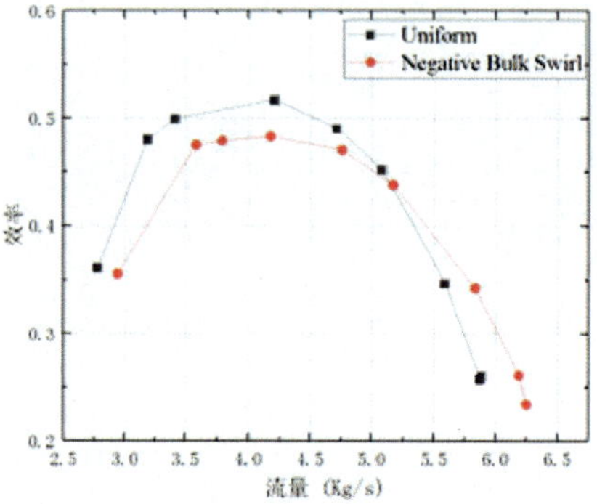

(b) Efficiency characteristics

Fig. 11. The effect of the overall swirl flow on the rotor characteristics (2400 rpm)

References

1. Walter OB, Anthony F (2012) Progress in boundary layer ingesting embedded engine research[C]. In: Joint propulsion conferences. American Institute of Aeronautics and Astronautics
2. Kawai RT, Friedman DM, Serrano L. Blended wing body (BWB) boundary layer ingestion (BLI) inlet configuration and system studies[R]. NASA contractor report, NASA/CR-2006-214534
3. Bealw DK, Cramer KB, King PS. Development of improved methods for simulating aircraft inlet distortion in turbine engine ground tests[R]. AIAA 2002-3045
4. 刘大响, 叶培梁, 胡骏, 等. 航空燃气涡轮发动机稳定性设计与评定技术 [M]. 北京 : 航空工业出版社 (2004)
5. 安玉戈, 刘火星. 压气机进气畸变数值模拟技术研究 [J]. 航空学报 (2012) 33(9):1624–1632
6. Stocks CP, Bissinger NC (1981) The design and development of the tornado engine air intake[R]. AGARD Aerodynamics of Power Plant Installation 21 p SEE N 82-1306504-01
7. Ludwig G (1989) Tomahawk engine/inlet compatibility study for F107-WR- 400/402 engines[R]. Williams international report CMEP-5003-2025
8. 屠宝锋, 胡骏. 旋流畸变影响低速轴流式压气机数值模拟研究 [J]. 推进技术 (2016) 37(9):1649–1656

9. Davis M, Beale D, Boyer K (2006) An example for integrated gas turbine engine testing and analysis using modeling and simulation[R]. ADA497076. ITEA
10. Milt D, Alan H (2007) A parametric study on the effects of inlet swirl on compression system performance and operability using numerical simulations[R]. GT2007-27033, ASME
11. Fredrick N, Milt D (2011) Investigation of the effect of inlet swirl on compressor performance and operability using a modified parallel compressor model[R]. GT2011-45553, ASME
12. Cousin W, Davis M (2011) Evaluating complex inlet distortion with a parallel compressor model: Part 1 concepts, theory, extensions, and limitations[R]. GT2011-45067, ASME
13. Sheoran Y, Bouldin B, Krishnan P (2012) Compressor performance and operability in swirl distortion. J Turbomach 134(4):2453–2464
14. 屠宝锋,胡骏,张凯,整体涡对跨声速单转子性能和稳定性影响[J].航空动力学报 (2016) 31(9):2233–2238
15. 屠宝锋,胡骏,张凯.对涡旋流影响压气机转子性能和稳定性的研究[J].推进技术 (2016) 37(4):641–645
16. 胡伟波, 程邦勤, 陈志敏, 等.整体涡旋流畸变对压气机性能影响的研究 [J]. 推进技术 (2015) 36(9) (Wei- bo H, Bang- qin C, Zhi-min C et al (2015) Investigation on Effects of Bulk Swirl Distortion on Compressor Performance [J]. J Propul Technol 36(9):1324–1330)
17. 程邦勤, 纪振伟, 胡伟波, 等.对涡旋流畸变对单级跨声速压气机性能影响的数值研究 [J]. 推进技术 (2017) 38(8):1776–1786 (Bang-qin C, Zhen-wei J, Wei-bo H et al (2017) Effects of paired swirl distortion on single stage transonic compressor performance using numerical simulation[J]. J Propul Technol 38(8):1776–1786)
18. 张磊, 程邦勤, 王加乐, 等. 新型旋流畸变网的设计与仿真研究[J]. 推进技术. (2018) 39(9):2110–2120 (Lei Z, Bangqin C, Jiale W et al (2018) Design and numerical simulation of a new swirl distortion screen[J]. J Propul Technol 39(9):2110–2120)

Check for updates

Target Localization on Image-Guided Missile

Hengbo Qi[1], Simei Ji[1](✉), Junmin Zhao[2], Jiyu Nie[2], and Chenrui Nie[1]

[1] School of Aerospace Engineering, Beijing Institute of Technology, Beijing 100081, P.R. China
x3815@bit.edu.cn
[2] Xi'an Modern Control Technology Institute, Xi'an 710000, P.R. China

Abstract. For image-guided missile, it is important to localize the target and its surrounding area using the images taken by seeker. In this paper, we propose a target localization method that combines on-board IMU data and image information. This method does not require equipment such as GPS or optoelectronic platform, nor does it need prior knowledge or templates of the target. Firstly, a selection strategy is proposed to determine keyframes, which are used to solve the pose of the seeker camera and the 3D coordinate of the target in the camera coordinate system. Then nonlinear optimization is used to reduce the reprojection error and obtain the optimal solution of the seeker camera pose and the 3D coordinate of the target. Finally, the scale uncertainty of the seeker monocular camera is solved by combining the on-board IMU data with images to obtain the coordinate of the target in the shooter coordinate system. The proposed method is validated in the flight data of a certain type of image-guided missile, and a high-precision localization result is obtained at a long missile-target distance.

Keywords: Target Localization · Passive Ranging · Image-Guided Missile

1 Introduction

Target localization can be achieved by solving for both the missile pose and the relative position of the missile-target in two steps. Typically, the missile pose (position and attitude) can be obtained from ground station, GPS (Global Positioning System) or on-board IMU (Inertial Measurement Unit) data. With the known missile pose, target localization is equivalent to determining the missile-target distance and direction [1]. Distance measurement methods are mainly divided into active ranging method and passive ranging method. Active ranging method, as laser ranging and electromagnetic wave ranging, has the advantages of long range and high accuracy, but requires active signal emission, which not only increase the burden of the missile platform, but also significantly reduce the concealment and be susceptible to enemy interference. On the other hand, passive ranging method such as visible image method, is inexpensive, stealthy and adaptable for many platforms. With the improvement of image sensor sensitivity and imaging resolution, passive ranging method based on visible image is receiving more and more attention [2].

Image-guided missiles are widely equipped in the military, which observe the battlefield, search and lock onto the target by on-board seeker. At the same time, the seeker

H. Ding (ed.), *Aerospace Mechatronics and Control Technology*, Springer Aerospace Technology,
https://doi.org/10.1007/978-981-16-6640-7_12

transmits abundant battlefield image information back to the ground launch vehicle through optical fiber. If this image information can be used to solve for the position of the target and its surrounding area, the attack effect can be improved in many ways, such as designing optimal guidance law that satisfies the fall point constraint and fall angle constraint [3], assisting in developing cluster attack strategy, or enhancing battlefield situational awareness. Therefore, how to quickly and accurately locate target and its surrounding area using on-board image information in the battlefield has become the focus of current research.

In the rest of this paper, the structure is as follows. Section 2 provides an overview of research on visible-image-based target localization methods. Then Sect. 3 presents the framework of our method in detail, i.e. how to localize target with image and on-board IMU data. Subsequently, Sect. 4 provides the result of performance of our method on the flight data, which is gathered by a certain type of image-guided missile. Finally, a brief conclusion and discussion are given in Sect. 5.

2 Related Work

The image sequence method is a commonly used passive ranging method, and the basic principle of it is to extract the features (such as grayscale, boundary, area, location and other obvious features) from the target and infer the missile-target distance from the change of features between adjacent image frames. In earlier work, Wu [4] and Fan [5] used the principle of pin-hole imaging to infer the change of missile-target distance from the change of the target imaging area between adjacent frames, and finally solved the missile-target distance at any flight process moment by iterative method, or the known true size of the target. However, this method is difficult to be applied to long-distance localization, especially when the target imaging area is small and susceptible to sub-pixel error, resulting in poor result. Wang [6] and Yu [7] proposed to use the "feature linearity" to describe the target. The feature linearity is visible on both adjacent frames, has 3D rotational invariance, and is only related to the missile-target distance. The change of the feature linearity can be used to deduce the change of the missile-target distance, but this method is difficult to be implemented on targets with insignificant linearity features. Xu [8] photographed the tank in all directions beforehand and saved the feature points extracted from the tank images as templates. Then this method compared the tank images captured by seeker with the saved templates and matched the feature points to estimate the position and attitude of the target tank at the same time. However, this method required a long preparation time and a large computational cost, resulting in its low generality.

In addition to image sequence method, Tao [9] firstly extracted the optical flow of the target on the image, and then derived an explicit expression of missile-target distance and the optical flow by the relation of camera motion and the change of the optical flow. However, this method required that the target be located in the center of the image and the feature points of the target are distributed on the same plane.

Zhou [10] and Cai [11] proposed a target localization method utilizing the photo-electric imaging platform and GPS. Firstly, GPS was used to obtain the position and attitude of the airborne optoelectronic platform, and the information such as azimuth angle and height angle of target is measured by optoelectronic platform. Then the target localization geometric model is established, which is able to solve multi-target location simultaneously according to the target imaging in the pixel plane. However, there is no equipment such as optoelectronic platform and GPS on an image-guided missile.

3 Proposed Method

To address the problems and limitations of the above target localization methods, we propose a method which does not require GPS or optoelectronic platform, nor does it need the prior knowledge or templates of target, and can simultaneously localize the target and its surrounding area by using only the on-board IMU data and the image information gathered by seeker. Firstly, keyframes are selected from a period of time based on a selection strategy, which are used to solve camera poses by epipolar constraint and PnP method, and target location by triangulation [12]. Then, the solved camera poses and target position are optimized using nonlinear optimization to reduce the reprojection error caused by computational and observation errors [13]. Finally, the scale uncertainty of the seeker monocular camera is solved by utilizing the on-board IMU data.

3.1 Coordinate Systems

Our method involves three coordinate systems: shooter coordinate system (w), missile-body coordinate system (b) and camera coordinate system (c), and the relationship between the three coordinate systems is shown in Fig. 1.

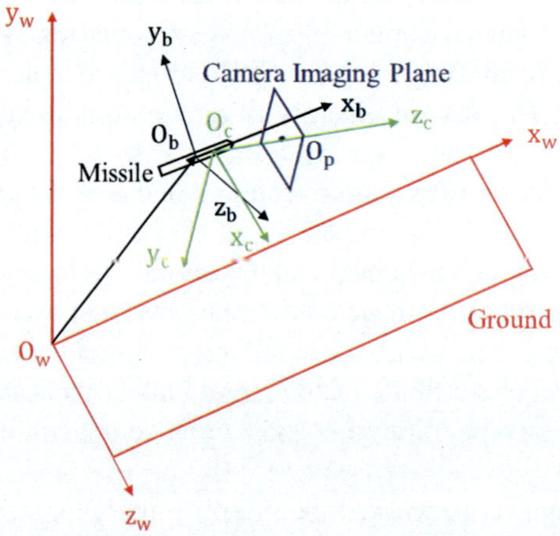

Fig. 1. Shooter, missile-body and camera coordinate system

The shooter coordinate system is inertial reference system, its origin O_w is the missile launch vehicle, $O_w y_w$ is perpendicular to the ground in upward direction, $O_w x_w$ is perpendicular to $O_w y_w$, facing the same direction as the missile launcher and $O_w z_w$ determined according to the right-hand rule.

The origin of the missile-body coordinate system O_b is located at the mass center of missile, O_bx_b faces to the head of missile, O_by_b is perpendicular to O_bx_b and lies in the missile's main symmetry plane, which coincides with the $x_wO_wy_w$ plane at the moment of missile launch. O_bz_b is determined according to the right-hand rule. The velocity and Euler angular velocity of the missile are measured by the on-board IMU equipment, and their integration yields the missile pose $H_{wb} \in \mathbb{R}^{4\times4}$ at each moment, expressed as formula (1).

$$
\begin{aligned}
H_{wb} &= \begin{bmatrix} R_{wb} & t_{wb} \\ 0 & 1 \end{bmatrix}, \\
R_{wb} &= \begin{bmatrix} \cos\theta\cos\psi & \sin\theta & -\cos\theta\sin\psi \\ -\sin\theta\cos\psi\cos\gamma + \sin\psi\sin\gamma & \cos\theta\cos\gamma & \sin\theta\sin\psi\cos\gamma + \cos\psi\sin\gamma \\ \sin\theta\cos\psi\sin\gamma + \sin\psi\cos\gamma & -\cos\theta\sin\gamma & -\sin\theta\sin\psi\sin\gamma + \cos\psi\cos\gamma \end{bmatrix}^T
\end{aligned}
\tag{1}
$$

where $\boldsymbol{t}_{wb} \in \mathbb{R}^{3\times1}$ and $R_{wb} \in \mathbb{R}^{3\times3}$ denote the position and attitude of the missile in the shooter coordinate system, respectively, and $R_{wb} \in \mathbb{R}^{3\times3}$ is determined according to three Euler angles (pitch angle ϑ, yaw angle ψ, roll angle γ).

The origin of camera coordinate system O_c is camera optical center, which usually does not coincide with the mass center of missile, and $\boldsymbol{t}_{bc} \in \mathbb{R}^{3\times1}$ is used to denote the relative transition between the camera optical center and the mass center of missile. O_cz_c passes through the geometric center O_p of the camera imaging plane, O_cO_p's length is the camera focal length f, O_cy_c aligns with the imaging plane longitudinally, and O_cx_c aligns with the imaging plane laterally. The camera intrinsic parameter is denoted by $K \in \mathbb{R}^{3\times3}$ (Fig. 2).

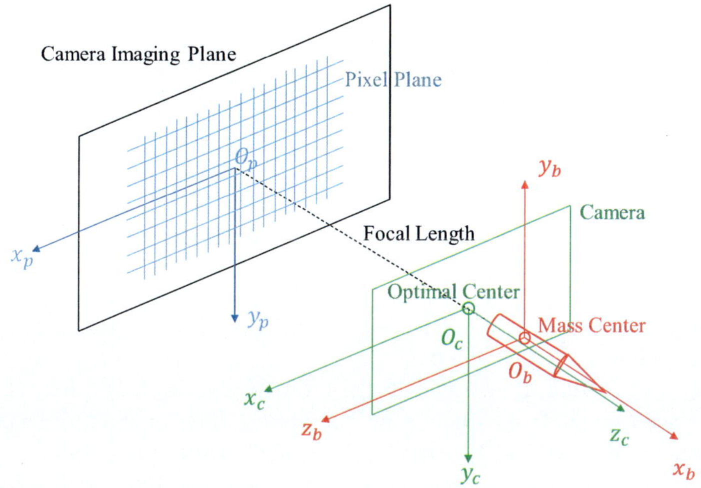

Fig. 2. Missile-body coordinate system, camera coordinate system and imaging plane

The rotation matrix between the camera coordinate system of seeker and the missile-body coordinate system is denoted by $R_{bc} \in \mathbb{R}^{3\times3}$, determined according to the yaw frame angle α and pitch frame angle β, with the positive direction of α being opposite to the positive direction of missile yaw angle ψ. $O_b x_b$ aligns with $O_c z_c$, $O_b z_b$ aligns with $O_c x_c$, and $O_b y_b$'s positive direction is opposite to $O_c y_c$, so R_{bc} is expressed as formula (2).

$$R_{cb}=\begin{bmatrix} 0 & 0 & 1 \\ 0 & -1 & 0 \\ 1 & 0 & 0 \end{bmatrix} * \begin{bmatrix} \cos\alpha\cos\beta & \sin\beta & \sin\alpha\cos\beta \\ -\cos\alpha\sin\beta & \cos\beta & -\sin\alpha\sin\beta \\ -\sin\alpha & 0 & \cos\alpha \end{bmatrix} \tag{2}$$

The transform relation of the camera coordinate system relative to the shooter coordinate system is expressed as formula (3).

$$R_{cw} = R_{cb} * R_{bw}, t_{cw} = R_{cb} * t_{bw} + t_{cb} \tag{3}$$

3.2 Target Localization Method

Our method is divided into the following four steps, and the flow diagram is shown in Fig. 3.

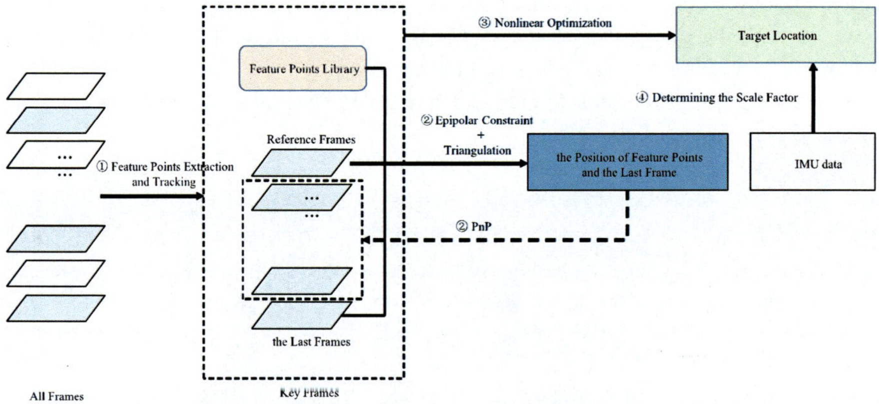

Fig. 3. Target localization method flow diagram

(1) Feature points extraction and tracking
Extract Fast feature points on selected reference frame, and track feature points on the subsequent frames using the LK optical flow method. Select keyframes according to the size and direction of the optical flows.
(2) Solving the keyframes' camera poses and the feature points' 3D coordinates
Basing on the observation of feature points on multiple frames, solve the keyframes' camera poses by epipolar constraint and PnP method, as well as the feature points' 3D coordinates by triangulation in the camera coordinate system of the reference frame.

(3) Nonlinear optimization
Project the solved 3D coordinates of the feature points to all keyframes (if the feature point is observed on that keyframe), and the pixel distance between its reprojection coordinate and its observation coordinate is called reprojection error. The reprojection error function is built and the keyframes' camera poses and feature points' 3D coordinates are optimized using nonlinear optimization method.
(4) Determining the scale factor
Align the camera motion trajectory solved by images with the motion trajectory obtained from the IMU data to determine the scale factor, and then add the scale factor to the keyframes' camera poses and the feature points' 3D coordinates. The pose of the reference frame in the shooter coordinate system can be obtained from the IMU data, so transform the keyframes' camera poses and the feature points' 3D coordinates to the shooter coordinate system according to formula (3). Finally, the target location is determined by finding the feature points of the target.

3.2.1 Feature Points Extraction and Tracking

A frame is selected as the reference frame during the flight, and its camera coordinate system rotation matrix is set to identity matrix I, and the coordinate is set to $(0, 0, 0)^T$. The subsequent frames' poses and feature points' coordinates are calculated in the reference frame camera coordinate system.

200 Fast feature points are extracted on the reference frame and are tracked using LK optical flow method on the subsequent frames. Because of the forward flight of the missile, the count of feature points will reduce due to the loss of optical flow tracking or the leaving of feature points from image area. A small count of feature points will result in large calculation error, so new feature points will be extracted to replenish the count of feature points on each frame to keep the count of feature points at 200. A feature point library is created to store all extracted feature points, recording the global number of each feature point, the numbers of all image frames on which it is observed and its pixel coordinates on those image frames.

Due to the high frequency of seeker camera, processing each image frame would require lots of computation cost. On the other hand, the optical flow between adjacent frames is too small to facilitate feature point triangulation. Therefore, a keyframe selection strategy based on the optical flow of feature points is proposed, and a frame is selected as a keyframe if the following four conditions are satisfied.

(1) the average optical flow length of all feature points is greater than 4 pixels.
(2) more than 80% feature points' optical flows move downward.
(3) after the previous keyframe, if none of the subsequent 4 frames is selected as a keyframe, then the 5th frame will be selected as a keyframe.
(4) the optical flow is accumulated from the reference frame, and the length of the accumulated optical flow of keyframe should be less than 60 pixels.

3.2.2 Solving the Keyframes' Camera Poses and the Feature Points' 3D Coordinates

After Sect. 3.2.1, the feature points jointly observed on the reference frame and the last frame are already known, as well as the pixel coordinates of these feature points on these two frames. Therefore, the rotation matrix $R_{c_1c_k}$ and translation $\boldsymbol{t}_{c_1c_k}$ of the last frame (number k) relative to the reference frame (number 1) are solved according to epipolar constraint.

In Fig. 4, when the camera observes the same landmark P at different viewpoints, the imaging of the landmark (pixel coordinates of the feature point) on these two frames will also be different, but satisfying the epipolar constraint of formula (4).

$$p_2^T F p_1 = 0 \tag{4}$$

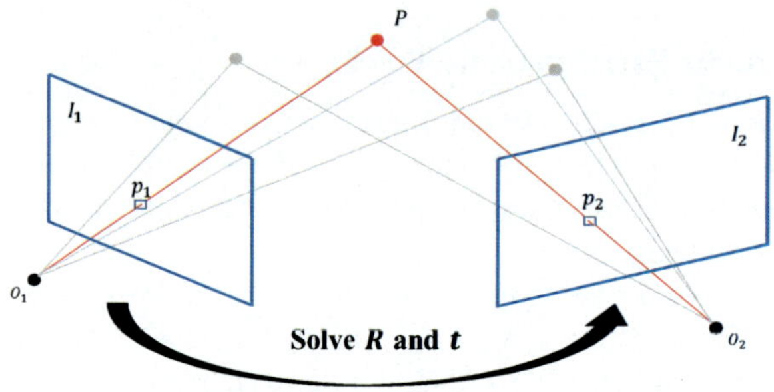

Fig. 4. Epipolar constraint

where $p_1, p_2 \in \mathbb{R}^{2\times 1}$ are the pixel coordinates of the landmark on the two frames respectively, and their 3×1 homogeneous coordinates are used here. $F \in \mathbb{R}^{3\times 3}$ represents the fundamental matrix of the epipolar constraint, which has 9 unknown parameters. But due to the monocular camera scale uncertainty, only eight equations are needed to solve for F. A constraint equation like formula (4) can be obtained from each landmark observation, and usually 8 jointly observed landmarks are needed. When there are more than 8 jointly observed landmarks between two image frames, the optimal solution is found by random sampling consistency (RANSAC).

$$F = K^{-T} \hat{t} R K^{-1} \tag{5}$$

The required $R_{c_1c_k}$ and $\boldsymbol{t}_{c_1c_k}$ can be solved by formula (5), where K is the camera intrinsic parameter and $\hat{\boldsymbol{t}}$ is the skew symmetric matrix of $\boldsymbol{t}$.

After the $R_{c_1c_k}$ and $\boldsymbol{t}_{c_1c_k}$ of the last frame relative to the reference frame are known, triangulation is used to solve for the 3D coordinates of the jointly observed landmarks. Figure 5 shows the triangulation method. When two cameras observe the same landmark, landmark's 3D coordinates can be estimated by its pixel coordinates. In Fig. 5, R_{c_1l} and $\boldsymbol{t}_{c_1l}$ are the pose of the last frame I_2 relative to the reference frame I_1, p_1 and p_2 are the pixel coordinates of the landmark on the two frames, s_1 and s_2 are the distances of the landmark P to the optical centers of the two cameras, respectively. In the left image, P must be located on the line of O_1p_1, and in the right image, P must be located on the line of O_2p_2, and the intersection point of the two lines is the 3D position of P, which satisfies the formula (6)

$$s_2p_2K^{-1} = s_1R_{c_1c_k}p_1K^{-1} + t_{c_1c_k} \tag{6}$$

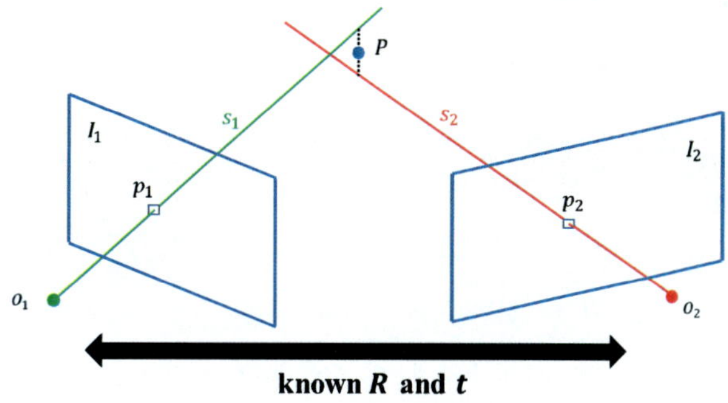

Fig. 5. Triangulation

Left multiply formula (6)'s both sides by $(p_2K^{-1}\hat{)}$

$$s_2(p_2K^{-1}\hat{)}(p_2K^{-1}) = s_1(p_2K^{-1}\hat{)}R_{c_1c_k}(p_1K^{-1}) + (p_2K^{-1}\hat{)}t_{c_1c_k} \tag{7}$$

The left of the formula (7) equals to 0, and the right is only related to s_1, therefore s_1 can be solved. So we can obtain the 3D coordinates of the landmark in the reference frame coordinate system. However, the two lines do not intersect strictly due to errors caused by calculation and observation, so there is a subsequent need to optimize the triangulation results.

At this point, the pose of the last frame relative to the reference frame is known, as well as the 3D coordinates of the landmarks jointly observed on these two frames, so the poses of other keyframes could be solved by PnP method (Fig. 6).

The PnP (Perspective-n-Point) method is used to solve for the camera pose based on the correspondence between the 3D structure and the 2D image when the 3D coordinates of the landmarks and their 2D pixel coordinates on an image frame are known. The commonly used PnP methods are P3P [14], EPnP (Efficient PnP) [15], and nonlinear optimization, etc. The EPnP method is used in this paper.

After solving the camera poses of all keyframes, the remaining feature points in the feature point library are triangulated.

3.2.3 Nonlinear Optimization

As stated above, there is reprojection error between the solved camera poses and the coordinates of the landmarks due to the calculation error and observation error, etc. Reprojection error is shown in Fig. 7.

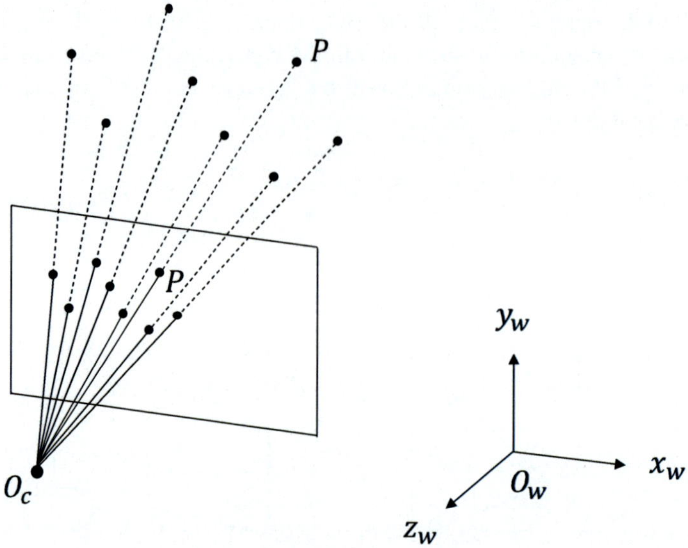

Fig. 6. PnP method

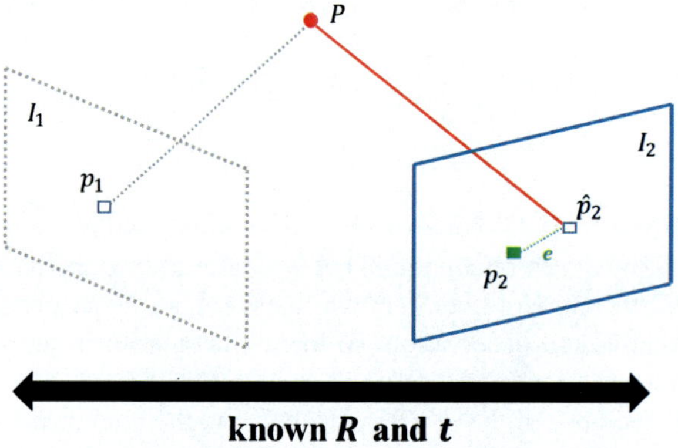

Fig. 7. Reprojection error

By projecting the 3D landmark P back to each keyframe (if it is observed on that frame), the difference e between the reprojected coordinate $\hat{p}_2$ and the observed coordinate p_2 of the landmark is called reprojection error. Let the pose of the i-th image frame in the reference frame camera coordinate system be $R_{c_ic_1}$ and $t_{c_ic_1}$, then the 3D coordinates of P in the i-th image frame camera coordinate system P' is

$$P' = R_{c_ic_0} * P + t_{c_ic_1} \tag{8}$$

The reprojection coordinate $\hat{p}_2$ is calculated by formula (9).

$$\hat{p}_2 = \left[\frac{KP'[1]}{P'[3]}, \frac{KP'[2]}{P'[3]} \right]^T \tag{9}$$

where K is the camera intrinsic parameter and $P'[\mathrm{m}]$ denotes the m-th dimension of $\hat{p}_2$. And the observation coordinate p_2 is known, then the reprojection error e is

$$e = \left\| \hat{p}_2 - p_2 \right\| \tag{10}$$

The number of landmarks is n and the number of keyframes is k. All reprojection errors are summed up, and all keyframe camera poses H and all landmarks' coordinates P are optimized using nonlinear optimization method to minimize the reprojection error.

$$(H, P)^* = \arg\min_{H,P} \sum_{j=0}^{n} \sum_{i=0}^{k} e_{ij} \tag{11}$$

3.2.4 Determining the Scale Factor

The images are used to solve for all the keyframe poses H and the coordinates of the landmarks P. However, because of the scale uncertainty of the seeker monocular camera, we cannot know the absolute position, for example, the position coordinate of the last frame is found to be $(1, 1, 1)^T$ in the reference frame camera coordinate system, but it cannot be determined whether the 1 here indicates 1 m, 5 m or 10 m in the shooter coordinate system, i.e., it can be scaled freely. Nevertheless, the seeker camera motion obtained from IMU data has absolute scale information, so the scale factor s can be determined by aligning the camera motion trajectories solved by the images and obtained from IMU data (Fig. 8).

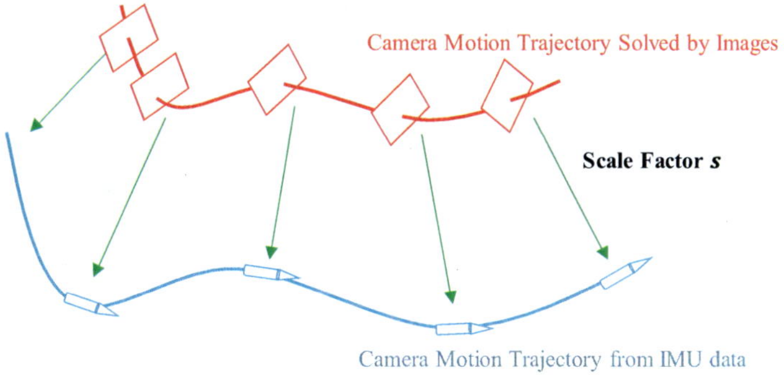

Fig. 8. Determining the scale factor

The upper right letter is specified to indicate the coordinate system where the variable is located in, c_1 is the reference frame camera coordinate system, and w is the shooter coordinate system. For two adjacent keyframes, let the first frame be i and the second frame be j. Then the translation A between image i and j is

$$A = (R_{c_1 b_i})^T * (T_{c_1 c_j}^{c_1} - T_{c_1 c_i}^{c_1}), R_{c_1 b_i} = R_{c_1 c_i} * R_{c_i b_i} \tag{12}$$

where, $R_{c_1 c_i}$ is the rotation matrix of i-th image frame relative to the reference frame, $R_{c_i b_i}$ is the rotation matrix of i-th image frame relative to missile-body, $R_{c_1 b_i}$ is the rotation matrix of i-th image frame's missile-body relative to the reference frame camera coordinate system, determined by multiplying $R_{c_1c_i}$ and $R_{c_ib_i}$. $T^{c_1}_{c_1c_i}$ represent the position of i-th image frame camera in the reference frame camera coordinate system.

The translation b obtained from the IMU data is

$$b = R_{c_1w} * (T^{w}_{wb_j} - T^{w}_{wb_i}) + (R_{c_1b_i})^T * R_{c_1b_j} * t_{bc} - t_{bc} \tag{13}$$

where R_{c_1w} is the rotation matrix of the reference frame camera coordinate system relative to the shooter coordinate system, $T^{w}_{wb_i}$ is the position of i-th image frame's missile-body in the shooter coordinate system, and $\boldsymbol{t}_{bc}$ is the relative translation of the camera optical center and the mass center of missile.

Therefore, the required scale factor s satisfies formula (14)

$$b = A * s \tag{14}$$

Left multiply formula (14)'s both sides by A^T, it becomes

$$A^T b = A^T A * s \tag{15}$$

There exists such a formula between every two frames, and for all keyframes, sum up all $A^T A$ and $A^T b$, then s could be determined.

$$\sum_{i=1}^{k-1} A^T b = \sum_{i=1}^{k-1} A^T A * s \tag{16}$$

The scale factor s will be multiplied into the keyframes' poses and landmarks' coordinates in the reference frame coordinate system. Finally, the reference frame camera pose R_{c_1w} and $\boldsymbol{t}_{c_1w}$ in the shooter coordinate system can be obtained from the IMU data, so all the results are transformed into the shooter coordinate system according to formula (4).

4 Results

The image-guided missile is launched at a missile-target distance of about 6000 m to attack a static target, and the frequency of both the on-board seeker camera and the IMU data is 50 Hz, and the image size is 1000 × 1000 pixel.

1000th frame is selected as the reference frame, and 14 keyframes are finally selected by the optical flow selection strategy: 1000, 1003, 1005, 1009, 1014, 1018, 1023, 1026, 1029, 1034, 1038, 1043, 1046, 1050. The optical flow tracking result is shown in Fig. 9. The white dots indicate the position of the feature point on the current frame, and the white lines are the size and direction of the optical flow, indicating the movement of the feature point from the previous frame to the current frame. The feature points in the lower area of the image are closer to the seeker, so the optical flows of these feature points are more obvious, while the optical flows of the feature points in the upper area are small. The target is located in the middle region of the image. When the missile-target distance is long, the target imaging area is small, and there is only one stable feature point on the target. According to the optical flow tracking result, the time histories of the target feature point pixel coordinates are shown in Fig. 10. In the image, the missile has locked onto the target and has been keeping the target in the center of the image, so the pixel coordinate change of the target feature point is small, and the slight jitter in the Y direction is caused by the image jitter during the missile flight.

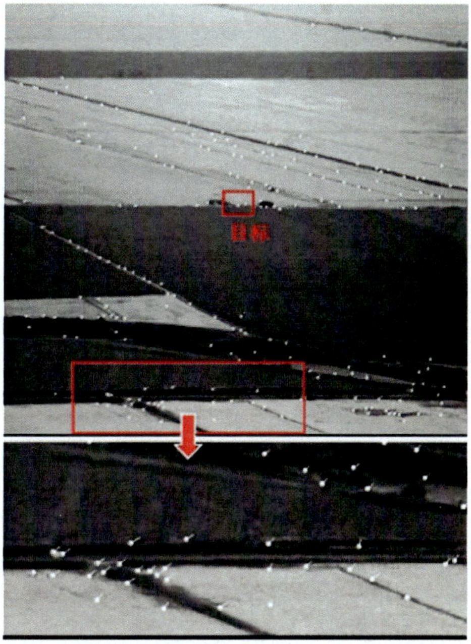

Fig. 9. Optical flow tracking result

The feature points jointly observed on the reference frame and the last keyframe are shown in Fig. 11. The white dots represent the positions of the feature points on the images. The feature points are extracted on the reference frame, while the feature points on the last keyframe are obtained by optical flow tracking, so the two feature points connected by a line segment actually correspond to a same landmark. The line segments are almost parallel, which indicates that the distribution of feature points on the reference frame and the last keyframe is the same, and the optical flow method tracks the feature points very well.

The reprojection coordinates before and after the nonlinear optimization are shown in Fig. 12. The light gray straight line in the image is a road, from which some feature points are extracted. The reprojection before optimization has deviated from the road, but after optimization it goes back to the road again, which shows that the reprojection error is reduced by optimizing the keyframe camera poses and the coordinates of landmarks.

Figure 13 shows the comparison of the camera motions solved by images and obtained from the IMU data. It can be seen that the motions are coincident in all three directions. There is small jitter in the Y direction and Z direction, which is caused by image jitter during the missile flight. 127.9 m transition in X direction, 1.31 m transition

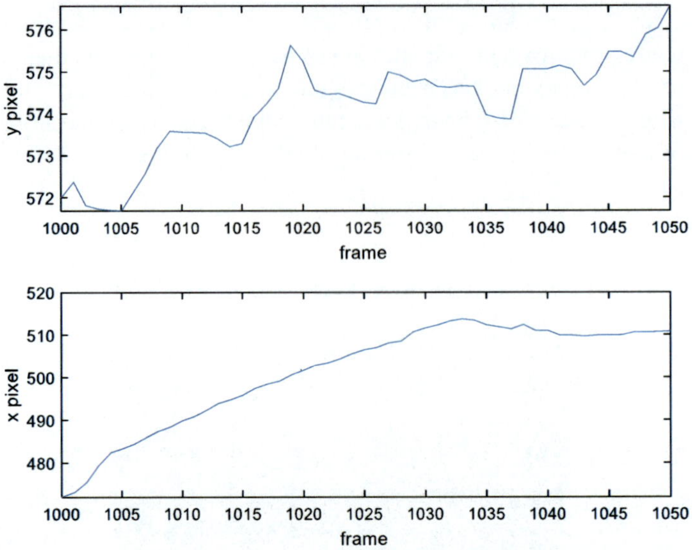

Fig. 10. The time histories of the target feature point pixel coordinates

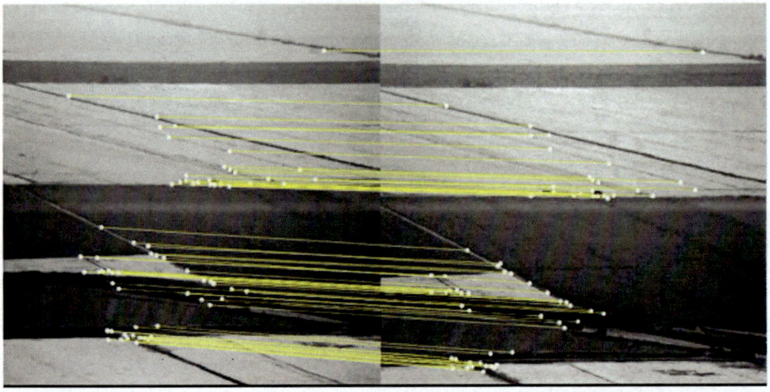

Fig. 11. The feature points jointly observed on the reference frame and the last frame

Fig. 12. The reprojection coordinates before (left) and after (right) the nonlinear optimization

in Y direction, and 12.1 m transition in Z direction, so the main motion is in X direction, and the jitter in the other two directions does not affect the results too much. Finally, the scale factor s is determined to be 112.3.

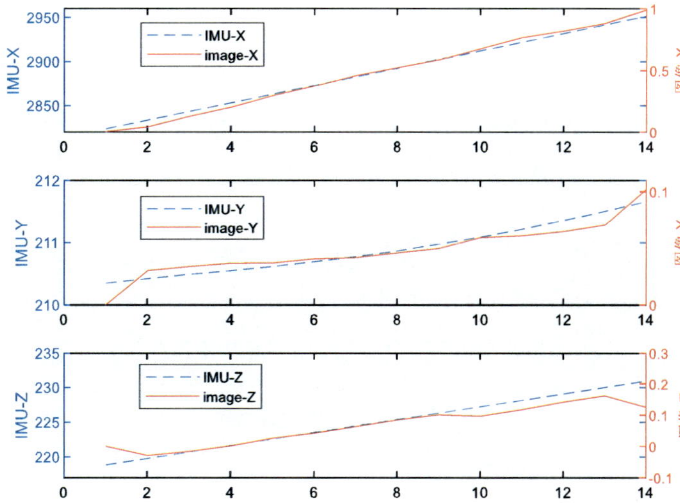

Fig. 13. Camera motions solved by images and obtained from the IMU data

The target localization result is shown in Table 1. With the missile-target distance at about 3390 m, the target localization error is only 93.38 m. The two main reasons for the error are as follows: (i) at the long missile-target distance, a small error in pose solution or landmark observation would greatly affect the localization results, though we have optimized it; (ii) there is a jitter in the image frame which results in inaccurate target pixel coordinate.

Table 1. The target localization result

	X(m)	**Y**(m)	**Z**(m)
Seeker position of the reference frame	2826.92	210.37	219.13
Real target location	6211.83	26.09	227.73
Estimated target location	6300.90	17.91	254.56
Error	89.07	8.18	26.83

The distribution of all the landmarks on the shooter coordinate system $O_w x_w z_w$ plane is shown in Fig. 14, which is consistent with the image observation, where the red dot is the target location. The landmarks in the lower area of Fig. 14 are closer to the missile, so they are denser. On the other hand, the landmarks in the left upper area are distributed discrete and separated by a large distance, whose distance from the missile are too far and the optical flows of their feature points are too small, leading to the inaccuracy triangulation results. Therefore, the localization results of these landmarks are less credible.

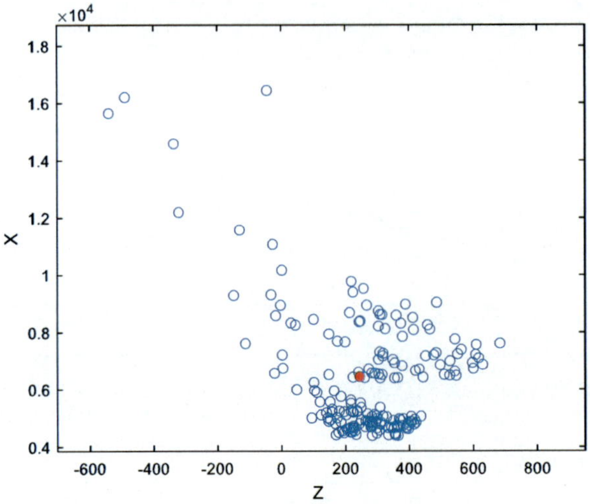

Fig. 14. The distribution of landmarks on $O_W x_W z_W$ plane

5 Conclusion

In this paper, a method was proposed to localize the target and its surrounding area only using the on-board IMU data and image data. The feature points were extracted on the reference frame and tracked to get the observation coordinates of the same landmark on each of keyframes, which were selected by optical flow. Then the seeker camera poses and the 3D coordinates of the landmarks were solved using the keyframes, and were optimized using nonlinear optimization to reduce the reprojection error. The scale factor was determined by aligning the camera motion trajectory obtained from the on-board IMU data with the one solved by the images. Finally, all the results were transformed from the reference frame camera coordinate system to the shooter coordinate system to complete the localization of the target. The method was validated on the flight data of a certain type of image-guided missile. The target localization error was only 93.38 m at the missile-target distance of 3390 m, and the distribution of landmarks was consistent with the image observation.

For the follow-up research, there are two prospects.

(1) The images taken by the seeker camera are not high definition and there are many textureless areas, which easily lead to the optical flow tracking error or lost, and finally cause the error of localization result. Nevertheless, there are obvious roads in the image from which feature lines can be extracted, so the combination of feature point and feature line can be used to improve the tracking result of features.

(2) This method considers all feature points to be static, but in the actual use of missiles, they usually attack dynamic targets, so it is necessary to conduct a research on the method of dynamic target localization. The dynamic target and static background in the image can be distinguished by classifying the optical flow, and then the position and motion of the dynamic target can be estimated after estimating the camera pose using the feature points on the static background.

References

1. Geng Z (2009) Research of key technology in passive ranging baed on image information. D. Xidian University (Chinese)
2. Han K, Desouza G (2009) Multiple targets geolocation using SIFT and Stereo vision on airborne video sequences, C. In: The 2009 IEEE/RSJ international conference on intelligent robots and systems, pp 5327–5332
3. Qiuqiu W, Qunli X, Chuntao C (2012) Method of range estimate based on relative position information of image seeker and INS. J Trans Beijing Inst Technol 32(02) 141–145(Chinese)
4. Hanping W (1994) Research on dynamic passive ranging method of multi-target in TV image guidance. J Modern Defense Technol 01: 43–49 (Chinese)
5. Minge F, Yan Z, Xiachuan S (2009) Missile-borne passive ranging of IR imaging based on area. J Infrared Laser Eng 38(03) (2009) 391–396(Chinese)
6. Di W (2011) Range measurement system and performance analysis based on monocular image and direction. J ACTA Photonica Sinica 40(07): 1112–1116(Chinese)
7. Yong Y (2009) Passive range-finding method of infrared imaging target based on characteristic lines. J Shipboard Electron Countermeas 32(6):86–89 (Chinese)
8. Chao X, Min G, Huan C (2015) The attitude angle estimation-based distance measurement of tank target in monocular image. J. ACTA Photonica Sinica 44(7):140–147 (Chinese)
9. Lin T (2005) The extraction of image information based on optical flow techniques. D. Huazhong University of Science and Technology (Chinese)
10. Qianfei Z, Jinghong L, Wenzhuo X, Yueming S (2015) Multi-target self-determination orientation system based on airborne photoelectric imaging platform. J ACTA Optica Sinica 25(01): 189–203(Chinese)
11. Mingbing C, Jinghong L, Fang X (2018) Multi- targets real- time location technology for UAV reconnaissance. J Chinese Optics 11(05): 812–821(Chinese)
12. Leutenegger S, Furgale P, Rabaud V et al (2013). Keyframe-based visual-inertial SLAM using nonlinear optimization, C/OL. In: Proceedings of robotics science and systems (RSS) (2013), pp 25–37
13. Qin T, Li P, Shen S (2018) VINS-Mono: a robust and versatile monocular visual-inertial state estimator. J IEEE Trans Robot 34(4):1004–1020
14. Xiaoshen G, Xiaorong H, Jianliang T et al (2003) Complete solution classification for the perspective-three-point problem. J IEEE Trans Pattern Analys Machine Intell 25(8) 930–943
15. Lepetit V, Moreno-Noguer F, Fua P (2009) EPnP: An accurate O(n) solution to the PnP problem. J Int J Comput Vision 81–87

Attitude Updating Algorithm Based on Agile Micro-Nano Satellites

Yajing Chen[1], Yun Fei[1], Tao Meng[1,2](✉), and Zhonghe Jin[1,2]

[1] School of Aeronautics and Astronautics, Zhejiang University, Hangzhou 310027, China
{Yer,fy927,mengtao,jinzh}@zju.edu.cn
[2] Zhejiang Micro-Nano Satellite Research Laboratory, Hangzhou 310027, China

Abstract. The high dynamic characteristics of agile micro-nano satellites will introduce noncommutativity error in the attitude updating process. Therefore, the attitude updating algorithm of the traditional micro-nano satellite is no longer applicable. This paper proposes a modified attitude updating scheme applied to agile satellites. The method adopts the coning motion as the research scene, in which the influence of noncommutativity error is the worst. Firstly, traditional attitude updating algorithm is presented, and based on the shortcoming of the algorithm, the rotation vector algorithm is introduced. Then, combined with the agile micro-nano satellite's characteristics, the upper bound and lower bound of the algorithm's sampling number are given. Besides, referring to the least-squares algorithm, we carry out angular velocity fitting and calculate the required equal interval angular velocity. Finally, the rotation vector compensation algorithm is derived to gain the updating attitude. The proposed algorithm does not need equal interval angular velocity sampling and can achieve higher estimation accuracy in a dynamic environment than the traditional algorithms. The simulation results validate the effectiveness of the proposed scheme.

1 Introduction

In recent years, the performance, reliability, and lifespan of micro-nano satellites have been improved by a wide margin. Therefore, micro-nano satellites have been successfully applied in various civil and military tactical applications [1]. With the increasing complexity of space missions, some micro-nano satellites are required to be able to maneuver rapidly around three axes [2]. These satellites are called agile micro-nano satellites. Compared with traditional micro-nano satellites, agile micro-nano satellites have characteristics of high precision pointing, fast multi-target acquisition and tracking. Besides, the attitude maneuver speed has been dadvanced by nearly one order of magnitude, which increases the complexity of the attitude calculation process [3].

According to Euler's rotation theorem, noncommutativity error will be introduced when integrating the dynamic angular velocity [4]. The angular velocity integral is involved in satellite attitude updating algorithms, so it is inevitable to introduce noncommutativity error. At present, the typical algorithms are the extended Karman filter (EKF) [5], the unscented Karman filter (UKF) [6], the cubature Kalman Filter (CKF) [7] and the particle filters (PFs) [8].

H. Ding (ed.), *Aerospace Mechatronics and Control Technology*, Springer Aerospace Technology,
https://doi.org/10.1007/978-981-16-6640-7_13

Traditional micro-nano satellites usually work in steady-state mode with slight velocity and demand relatively low attitude estimation accuracy. Hence, the noncommutativity error can be ignored. Unlike traditional micro-nano satellites, agile satellites are usually in high dynamic situations and require high attitude estimation accuracy. The impact of noncommutativity error is too high to ignore [9].

There is little research on c errors in the field of satellite, but there is much studying in the field of inertial navigation. Bortz proposed the equivalent rotation vector equation [10] in 1971, which could compensate for the noncommutativity error. On this foundation, Ben et al. derived a conic optimal attitude algorithm in the form of angular rate input [11], Chen et al. proposed an attitude algorithm based on the secondary optimization of the periodic term error compensation coefficient under angular rate input [12]. Tang et al. established an attitude optimization goal based on the least-squares principle and proposed a new conic optimization algorithm under angular rate input [13, 14].

However, in practice, the attitude updating frequency of micro-nano satellites is much lower than that of the inertial navigation system. Further, the gyroscope accuracy is lower. The above factors make the optimization algorithm not directly applicable in the satellite's attitude determination system. In this paper, combined with the specific problems in micro-nano satellites' engineering, an improved attitude updating algorithm for agile micro-nano satellites is proposed based on the above algorithms. Finally, through simulation, the correctness of the proposed algorithm is verified.

2 Traditional Attitude Updating Method

Whether utilizing angular velocity integral to update attitude directly or estimating by filtering algorithms, the essence is to solve the differential equation of kinematics:

$$\dot{\mathbf{q}} = \frac{1}{2}\mathbf{q} \otimes \boldsymbol{\omega} \tag{1}$$

where $\mathbf{q}$ denotes the attitude quaternion and $\boldsymbol{\omega}$ denotes the angular velocity of the satellite. The usual method is to use the one-order approximated Peano-Baker Quaternion algorithm to obtain the solution of the attitude quaternion differential equation as follows:

$$\mathbf{q}(t_{k+1}) = e^{\frac{1}{2}\int_{t_k}^{t_{k+1}} M(w)dt} \cdot \mathbf{q}(t_k) \tag{2}$$

Adopt notation as follows:

$$\begin{aligned} \Delta\Theta &= \int_{t_k}^{t_{k+1}} M(w)dt = \int_{t_k}^{t_{k+1}} \begin{bmatrix} 0 & -w_x & -w_y & -w_z \\ w_x & 0 & w_z & -w_y \\ w_y & -w_z & 0 & w_x \\ w_z & w_y & -w_x & 0 \end{bmatrix} dt \\ &= \begin{bmatrix} 0 & -\Delta\theta_x & -\Delta\theta_y & -\Delta\theta_z \\ \Delta\theta_x & 0 & \Delta\theta_z & -\Delta\theta_y \\ \Delta\theta_y & -\Delta\theta_z & 0 & \Delta\theta_x \\ \Delta\theta_z & \Delta\theta_y & -\Delta\theta_x & 0 \end{bmatrix} \end{aligned} \tag{3}$$

where $\Delta\theta$ denotes the attitude angle increment.

For the sake of simplicity, the most commonly used is the first-order approximation algorithm:

$$\mathbf{q}(t_{k+1}) = \Delta\Theta \cdot \mathbf{q}(t_k) \tag{4}$$

where $\mathbf{q}(t_k)$ denotes the attitude quaternion of the last moment.

Ordinarily, the integral process of Eq. (3) is approximately calculated, where the angular velocity is regarded as constant or first-order curve as follows:

$$\Delta\boldsymbol{\theta}= \int_{t}^{t+\Delta T} \mathbf{w}(\tau)\mathrm{d}\tau \approx \mathbf{w} \cdot \Delta T \approx \frac{\mathbf{w}_t + \mathbf{w}_{t+\Delta T}}{2} \cdot \Delta T \tag{5}$$

where $\mathbf{w}_i$ denotes the angular velocity of the satellite at i-time.

Traditional micro-nano satellites require low attitude determination precision. The variety of angular velocity is negligible, so the approximation error brought by Eq. (5) has little effect on the attitude estimation. The attitude accuracy requirements for agile micro-nano satellites are high, and large-angle fast maneuvers are required, leading to a complex angular velocity curve. Therefore, the approximate processing method in Eq. (5) will bring significant errors. It is necessary to perform sufficient high-precision fitting angular velocity.

Besides, since the agile micro-nano satellites don't rotate on a fixed axis when maneuvering, the direction is constantly changing in space. That is, the angular increments gained by Eq. (5) are inconsistent with the actual angular increments, and the noncommutativity error is introduced. In sum, the traditional Peano-Baker Quaternion algorithm is not suitable for agile micro-nano satellites.

3 Rotation Vector Algorithm

3.1 Rotation Vector Algorithm

To reduce noncommutativity error, the rotation vector equation proposed by Bortz is used to update the attitude, which can be expressed as:

$$\dot{\boldsymbol{\Phi}}=\mathbf{w} + \frac{1}{2}\boldsymbol{\Phi} \times \mathbf{w} + \frac{1}{|\boldsymbol{\Phi}|^2}\left[1 - \frac{|\boldsymbol{\Phi}|\sin|\boldsymbol{\Phi}|}{2(1-\cos|\boldsymbol{\Phi}|)}\right]\boldsymbol{\Phi} \times (\boldsymbol{\Phi} \times \mathbf{w}) \tag{6}$$

where $\mathbf{w}$ denotes the angular velocity of the satellite, $\boldsymbol{\Phi}$ denotes the rotation vector. The sum of the last two terms on the right side of the equation represents the noncommutativity error.

In the application of practical engineering, one takes the first two terms to guarantee real-time and the convenience of calculation:

$$\dot{\boldsymbol{\Phi}}=\mathbf{w} + \frac{1}{2}\boldsymbol{\Phi} \times \mathbf{w} \tag{7}$$

As can be seen from Eq. (7), when the direction of the angular velocity remains invariable, the Euler axis of rotation is parallel to the direction of angular momentum. The vector product of the rotation vector and the angular velocity is zero, and the incremental angle calculated by Eq. (5) is nearly error-free at this time. When the direction of the angular velocity changes, the vector product is nonzero, so the angular increment directly calculated by Eq. (5) has a missing term relative to the actual angular increment, which is noncommutativity error.

When the Euler axis of rotation is perpendicular to the angular velocity, the introduced noncommutativity error is the maximum. This kind of motion scene is the coning motion. Suppose there is a coning motion rotating around the x-axis of the inertial coordinate system:

$$\mathbf{\Phi}=\begin{bmatrix}0 & \alpha\cos(\Omega t) & \alpha\sin(\Omega t)\end{bmatrix} \tag{8}$$

where α is the half cone angle, Ω denotes the angular frequency. The corresponding angular velocity is:

$$\mathbf{w}=\begin{bmatrix}-2\Omega\sin^2(\alpha/2)\\ -\Omega\sin\alpha\sin(\Omega t)\\ \Omega\sin\alpha\cos(\Omega t)\end{bmatrix} \tag{9}$$

3.2 Coning Error Compensation Algorithm

According to Eq. (8), the quaternion of coning motion is:

$$\mathbf{q}(t)=\begin{bmatrix}0 & \sin\frac{\alpha}{2}\cos(\Omega t) & \sin\frac{\alpha}{2}\sin(\Omega t) & \cos\frac{\alpha}{2}\end{bmatrix} \tag{10}$$

It can be drawn to the theoretical quaternion increment from time t to (t + T) as follows:

$$\Delta\mathbf{q}=\begin{bmatrix}-\sin^2\frac{\alpha}{2}\sin(\Omega T)\\ -\sin\alpha\sin\frac{\Omega T}{2}\sin\left[\Omega\left(t+\frac{T}{2}\right)\right]\\ \sin\alpha\sin\frac{\Omega T}{2}\cos\left[\Omega\left(t+\frac{T}{2}\right)\right]\\ 1-2\sin^2\frac{\alpha}{2}\sin^2\frac{\Omega T}{2}\end{bmatrix} \tag{11}$$

When the Euler angle is tiny, one obtains the corresponding rotation vector increment as follows:

$$\Delta\mathbf{\Phi}=\begin{bmatrix}-2\sin^2\frac{\alpha}{2}\sin(\Omega T)\\ -2\sin\alpha\sin\frac{\Omega T}{2}\sin\left[\Omega\left(t+\frac{T}{2}\right)\right]\\ 2\sin\alpha\sin\frac{\Omega T}{2}\cos\left[\Omega\left(t+\frac{T}{2}\right)\right]\end{bmatrix} \tag{12}$$

Integrating the theoretical angular velocity directly according to Eq. (9), we get the angular increment within a period as follows:

$$\Delta\mathbf{\theta}=\int_t^{t+T} w(\tau)\mathrm{d}\tau=\begin{bmatrix}-2\Omega T\sin^2\frac{\alpha}{2}\\ -2\sin\alpha\sin\frac{\Omega T}{2}\sin\left[\Omega\left(t+\frac{T}{2}\right)\right]\\ 2\sin\alpha\sin\frac{\Omega T}{2}\cos\left[\Omega\left(t+\frac{T}{2}\right)\right]\end{bmatrix} \tag{13}$$

Comparing Eq. (12) with Eq. (13), one can notice that the second and third components are precisely the same and periodically change. But there is a constant deviation between the first item, which is equivalent to a constant angular velocity integral deviation. Therefore, the compensation algorithm aims at minimizing the constant error.

Assuming that the initial time is t, the angular velocity of each sampling point is:

$$\mathbf{w}_i = \begin{bmatrix} -2\Omega\sin^2(\alpha/2) \\ -\Omega\sin\alpha\sin\left(\Omega\left(t+\frac{i-1}{n-1}T\right)\right) \\ \Omega\sin\alpha\cos\left(\Omega\left(t+\frac{i-1}{n-1}T\right)\right) \end{bmatrix}, i = 1, 2, v, n \tag{14}$$

The following relationship can be found from each sampling value of the angular velocity of each point:

$$\mathbf{w}_i \times \mathbf{w}_j = \begin{bmatrix} \Omega^2\sin^2\alpha\sin\left(\frac{j-i}{n-1}\Omega T\right) \\ 2\Omega^2\sin\alpha\sin^2\frac{\alpha}{2}\left\{\cos\left(\Omega\left(t+\frac{i-1}{n-1}T\right)\right)-\cos\left(\Omega\left(t+\frac{j-1}{n-1}T\right)\right)\right\} \\ 2\Omega^2\sin\alpha\sin^2\frac{\alpha}{2}\left\{\sin\left(\Omega\left(t+\frac{i-1}{n-1}T\right)\right)-\sin\left(\Omega\left(t+\frac{j-1}{n-1}T\right)\right)\right\} \end{bmatrix} \tag{15}$$

It can be seen from the above equation that in the cross-product result of the sampling point, the first term is constant. And the second and third terms change periodically, whose integral results are close to zero. Furthermore, the constant term contains the period T, and its value varies with the sampling data. Hence, the vector product of the angular velocity at different sampling point can be used to compensate for the noncommutativity error:

$$\Delta\hat{\Phi}=\Delta\hat{\theta}+T^2\sum_{i=1}^{n-1}K_i\mathbf{w}_1\times\mathbf{w}_{i+1} \tag{16}$$

where K_i is the compensation coefficient.

The optimization objective is to choose appropriate K_i which minimizes $\Delta\boldsymbol{\Phi}_x-\Delta\hat{\boldsymbol{\Phi}}_x$:

$$\begin{aligned}\Delta\boldsymbol{\Phi}_x-\Delta\hat{\boldsymbol{\Phi}}_x &= -2\sin^2\frac{\alpha}{2}\sin(\Omega T)+2\Omega T\sin^2\frac{\alpha}{2}-T^2\Omega^2\sin^2\alpha\sum_{i=1}^{n-1}K_i\sin\left(\frac{i}{n-1}\Omega T\right) \\ &= 2\sin^2\frac{\alpha}{2}\sum_{k=2}^{\infty}(-1)^{k-1}\frac{(\Omega T)^{2k-1}}{(2k-1)!}-T^2\Omega^2\sin^2\alpha\sum_{i=1}^{n-1}K_i\sin\left(\frac{i}{n-1}\Omega T\right)\end{aligned} \tag{17}$$

Equation (17) can be simplified as:

$$\begin{aligned}\sum_{k=2}^{\infty}(-1)^{k-1}\frac{(\Omega T)^{2k-1}}{(2k-1)!} &= 2T^2\Omega^2\sum_{i=1}^{n-1}K_i\sum_{k=1}^{\infty}(-1)^{k-1}\frac{(\Omega T)^{2k-1}}{(2k-1)!}\left(\frac{i}{n-1}\right)^{2k-1} \\ &= 2\sum_{i=1}^{n-1}K_i\sum_{k=1}^{\infty}(-1)^{k-1}\frac{(\Omega T)^{2k+1}}{(2k-1)!}\left(\frac{i}{n-1}\right)^{2k-1}\end{aligned} \tag{18}$$

To facilitate the account, reformulate Eq. (18) to matrix form as:

$$\begin{aligned}&\left[(\Omega T)^3\ (\Omega T)^5\ (\Omega T)^7\ \cdots\right]\begin{bmatrix}C_1\\C_2\\C_3\\\vdots\end{bmatrix}\\&=\left[(\Omega T)^3\ (\Omega T)^5\ (\Omega T)^7\ \cdots\right]\begin{bmatrix}A_{11}\ A_{12}\ \cdots\ A_{1(n-1)}\\A_{21}\ \ddots\ \vdots\ A_{2(n-1)}\\ A_{pq}\\\vdots\ \ \vdots\ \ \vdots\end{bmatrix}\begin{bmatrix}K_1\\K_2\\\vdots\\K_{n-1}\end{bmatrix}\end{aligned}\tag{19}$$

where $C_i = \frac{(-1)^{i-1}}{(2i+1)!}, i = 1, 2, 3\cdots; A_{pq} = 2\frac{(-1)^{p-1}}{(2p-1)!}\left(\frac{q}{n-1}\right)^{2p-1}, q = 1 \sim n-1, p = 1, 2, 3\cdots$

The value of K_i can be gained by ensuring the first (n-1) terms of AK and C are the same. This algorithm is called the n-sample rotation vector attitude algorithm under angular rate input. Generally speaking, the more samples we obtained, the higher the compensation accuracy is. When the period is small, the compensation accuracy can be expressed as:

$$\varepsilon=\frac{1}{2}\alpha^2\left(C_n-\sum_{j=1}^{n-1}A_{nj}K_j\right)(\Omega T)^{2n+1}\tag{20}$$

4 Modified Attitude Updating Algorithm

The derivation of the rotation vector method is based on the theoretical angular velocity function. However, in practice, we can only get the gyro sampling data other than the theoretical value. The sampling angular velocity contains noise, which limits the compensation accuracy of the rotation vector attitude algorithm. When the number of samples reaches a certain value, the compensation accuracy will not be improved anymore. Consequently, the number of samples should be analyzed combining with the actual gyro instead of increasing blindly.

In the light of Eq. (16), gyro noise is introduced into the equation in the form of vector product. No matter how many samples are, the magnitude of compensation accuracy can not be less than that of noise. According to this, the upper bound of the number of samples can be approximately selected as follows:

$$v_g^2 \le \frac{1}{2}\alpha^2\left(C_n-\sum_{j=1}^{n-1}A_{nj}K_j\right)(\Omega T)^{2n+1}\tag{21}$$

where v_g is gyro random noise.

EKF is the most prevalent algorithm in satellite attitude estimation, in which gyro bias is usually chosen as another estimated parameter. From EKF filtering theory, the bias between $\Delta\Phi_x$ and $\Delta\theta_x$ is considered to be part of the gyro bias. As a consequence, the influence of the noncommutativity error on attitude angle updating can be neglected. Instead, it is mainly reflected in the bias estimation, so the lower bound of the number of samples can be defined as:

$$\tilde{b} \geq \frac{1}{2}\alpha^2 \left(C_n - \sum_{j=1}^{n-1} A_{nj} K_j \right) (\Omega T)^{2n+1} \tag{22}$$

where $\tilde{b}$ is the estimation accuracy index of gyro bias.

It can be found from Eq. (13) and Eq. (16) that the theoretical angular velocity function is used in the derivation. Nevertheless, we can only fit the angular velocity curve with samples in practice. Both gyro noise and the number of samples will affect the fitting accuracy. The n-sample rotation vector attitude algorithm obtains n new equal interval sampling values every period. Assuming that four-sample algorithm is adopted, the fourth-order angular velocity curve can be fitted:

$$\mathbf{w}(t) = a + 2bt + 3ct^2 + 4dt^3 + 5et^4 \tag{23}$$

The parameters can be calculated by substituting the sampling values into Eq. (23):

$$\begin{cases} a = \mathbf{w}_{t=0} \\ b = \dfrac{1}{6T}\left(-25\mathbf{w}_{t=0} + 48\mathbf{w}_{t=T/4} - 36\mathbf{w}_{t=T/2} + 16\mathbf{w}_{t=3T/4} - 3\mathbf{w}_{t=T}\right) \\ c = \dfrac{1}{9T^2}\left(70\mathbf{w}_{t=0} - 208\mathbf{w}_{t=T/4} + 228\mathbf{w}_{t=T/2} - 112\mathbf{w}_{t=3T/4} + 22\mathbf{w}_{t=T}\right) \\ d = \dfrac{1}{3T^3}\left(-20\mathbf{w}_{t=0} + 72\mathbf{w}_{t=T/4} - 96\mathbf{w}_{t=T/2} + 56\mathbf{w}_{t=3T/4} - 12\mathbf{w}_{t=T}\right) \\ e = \dfrac{1}{15T^4}\left(32\mathbf{w}_{t=0} - 128\mathbf{w}_{t=T/4} + 192\mathbf{w}_{t=T/2} - 128\mathbf{w}_{t=3T/4} + 32\mathbf{w}_{t=T}\right) \end{cases} \tag{24}$$

One gets the angular increment by integrating Eq. (23):

$$\begin{aligned} \hat{\theta} &= aT + bT^2 + cT^3 + dT^4 + eT^5 \\ &= \frac{T}{90}\left(7\mathbf{w}_{t=0} + 32\mathbf{w}_{t=T/4} + 12\mathbf{w}_{t=T/2} + 32\mathbf{w}_{t=3T/4} + 7\mathbf{w}_{t=T}\right) \end{aligned} \tag{25}$$

In fact, it is possible to get angular velocity at non equidistant sampling points, which leads to the unavailability of Eq. (16) and Eq. (25). Considering the timestamp of sampling points can be obtained at the same time, therefore, the angular velocity can be described by m-order curve:

$$\mathbf{w}_i = \left[\, 1 \; t_i^1 \; \cdots \; t_i^m \,\right] X + \mathbf{v}_i \tag{26}$$

where $\mathbf{w}_i$ is the ith sampling angular velocity;t_i denotes the time of the ith sampling value;$\mathbf{v}_i$ is the gyro's noise; X is the (m + 1)-dimensional vector and represents the fitted coefficient.

When there are n(n ≥ m) sample values during one period, we can estimate the fitted coefficient by the least square algorithm:

$$\hat{X} = \left(T^T T\right)^{-1} T^T W \tag{27}$$

where $T = \begin{bmatrix} 1 \; t_0^1 \cdots t_0^m \\ \vdots \; \vdots \; \ddots \; \vdots \\ 1 \; t_n^1 \cdots t_n^m \end{bmatrix}$, $W = \begin{bmatrix} w_0 \; w_1 \; \cdots \; w_n \end{bmatrix}^T$

Compared with the equal interval fitting method, the least square fitting method's disadvantage is that the coefficients need to be recalculated every period. But its merits are: There is no need for equal interval sampling; the equal interval angular velocity can be calculated, and Eq. (16) is still applicable; the gyro's noise can be reduced to a certain extent.

In sum up, the modified attitude updating algorithm for agile micro-nano satellites can be described as follows, which is called modified n-sample algorithm:

(1) Combining with the gyro's parameters and precision requirements for gyro bias, calculate the upper and lower bounds of the number of samples according to Eq. (21) and Eq. (22). Then choose the appropriate n-sample algorithm and find compensation coefficient *K*;
(2) Sample at least n data, and fit the angular velocity curve of the current period;
(3) Calculate n equal interval angular velocities from the fitted angular velocity function;
(4) Update attitude according to Eq. (16).

5 Simulation

Considering an agile micro-nano satellite equipped with a gyroscope and a star sensor, the EKF algorithm based on MRPs is used as the attitude estimation algorithm. The relevant parameters are shown in Table 1.

Table 1. Simulation parameters

Parameter	Description	Value
Δt	Simulation step	1 s
J	The inertia moment of the satellite	diag([1.2 1.2 1.5])kg.m^2
vg	Measurement accuracy of gyro	0.003°/s
b0	Initial bias of the gyro	[0.2 −0.2 0.5]^T°/s
vs	The noise of the star sensor	[50 10 10]^T arcsecond

Set the coning motion angle frequency $\Omega = 2$ Hz, and half coning angle $\alpha = 2°$ in the light of Eq. (9). Precision requirement for gyro bias is ten arcsec per second. Then we can calculate that the lower bound of the number of samples is two-sample, and the upper bound is three-sample based on Eqs. (21), (22).

The conventional algorithm (that is, one-order Peano-Baker Quaternion algorithm), the traditional two-sample algorithm, and the modified two-sample algorithm are used for comparison. For the sake of improving the fitting accuracy of angular velocity, the modified two-sample algorithm samples five new angular velocities in each period. The root mean square error of estimation error after convergence is summarized in Table 2.

Table 2. Accuracy comparison of compensation algorithms

Algorithm	Attitude angle estimation accuracy($''$)			Gyro bias estimation accuracy($''$/s)		
	Roll angle	Pitch angle	Yaw angle	Roll axis	Pitch axis	Yaw axis
Conventional	336.08	939.13	939.79	428.87	60.64	82.77
Traditional two-sample	11.42	6.29	6.28	128.06	1.24	1.70
Modified two-sample	9.27	3.12	3.09	1.78	1.01	1.63

From Table 2, the results report that the conventional algorithm is not applicable to the coning motion, whose bias estimation accuracy is one order of magnitude lower than the index. Compared with the conventional algorithm, the traditional two-sample accuracy is increased by four times, but it is still not meeting the requirement. By contrast, the modified two-sample algorithm uses six samples to fit the angular velocity curve and gets two new equispaced angular velocities, which improves the fitting accuracy and indirectly reduces the gyro's noise, so the accuracy is better and satisfies the demand. These findings suggest that the proposed algorithm is more effective.Simultaneously, to verify the upper and lower bound formulas, the following types of simulations are considered: single-sample algorithm, two-sample algorithm, three-sample algorithm, and four-sample algorithm. The table below illustrates the algorithms' RMS after convergence. It can be seen that the accuracy of the roll axis's bias estimation of the single-sample algorithm and the four-sample algorithm does not meet the target. The principal reason is that the single-sample does not compensate for the noncommutativity error, which only carries out the curve fitting. According to the coning motion parameters setting, we can calculate that the difference of $\Delta\Phi_x$ and $\Delta\hat{\Phi}_x$ is about $137''$ in the single-sample. Since the attitude determination period is 1 s, the noncommutativity error is equivalent to introducing $137''$/s in the roll axis's gyro bias, corresponding to the results in Table. 3. And the decrease of bias accuracy in the three-sample algorithm is mainly caused by the gyro noise. That is because the three-sample method is the upper bound's critical value, and the upper bound calculation formula is approximate, so it may also be affected by noise. It is apparent from this table that the two-sample algorithm is the most suitable, implying the method to determine the upper bound and lower bound is correct.

Table 3. Accuracy comparison of modified algorithms for different samples

Algorithm	Attitude angle estimation accuracy($''$)			Gyro bias estimation accuracy($''$/s)		
	Roll angle	Pitch angle	Yaw angle	Roll axis	Pitch axis	Yaw axis
Single sample	9.01	3.06	3.07	136.93	0.83	1.85
Two sample	9.00	3.03	3.05	1.06	0.82	1.84
Three sample	9.00	3.06	3.05	27.99	0.82	1.83
Four sample	9.00	3.07	3.05	43.37	0.82	1.83

6 Conclusion

This paper focuses on attitude updating of agile micro-nano satellites. Firstly, the shortcomings of traditional algorithms are analyzed, implying that angular velocity fitting accuracy and noncommutativity error must be considered. On this basis, an modified attitude updating scheme is proposed. The simulations representing the proposed method's bias estimation accuracy are two orders of magnitude higher than the traditional algorithms. This research provides a timely and necessary study of the agile micro-nano satellites' attitude determination.

Acknowledgements. This work was supported by Primary Research and Development Plan of Zhejiang Province, key technologies of Nano remote sensing satellites, 209C05004; Zhejiang Provincial Key Laboratory of Micro-nano Satellite.

References

1. Li JQ, Wang QY, Zhao SZ (2020) Spacecraft engineering 29(04):126–132
2. Baoyin HY (2020) J Dyn Control Syst 18(04):1–11
3. Zhao X, Wang SC, Yang DF, Luo YG (2011) J Chin Inertial Technol 19(02):163–169
4. Zhi X, Jianye L, Yanping W (2003) J Southeast U: Nat Sci Ed 33(04):419–423
5. Wang S, Zhang S, Zhou B, Wang X (2019) IEEE Access 7:20060–20075
6. Lee D, Vukovich G, Lee R (2017) J Aerospace Eng 30(4)
7. Yuan X, Zhang C, Shi C (2017) The Cubature Kalman Filter and its application in the satellite star-sensor/gyro attitude determination system. In: The 8th China Satellite Navigation Conference, 1. Shanghai, China
8. XiaoLei H, CongZhe Z, Yong L, YiLan PQ (2020) J Astronautics 41(08):1032–1041
9. Yin J, Meng Y, Zhang WL, Hong C (2016) Comput Simulat 33(07):147–152+174
10. Bortz JE (1971) IEEE T Aero Elec Syst AES-7(1):61–66
11. Ben Y, Sun F, Yu, Wei GF (2009) IEEE T Aero Elec Syst 45(4):1565–1572

12. Chen JF, Zhu XC (2012) J Chinese Inertial Technol 20(02):131–135
13. Tang CY, Li XC (2013) J Chinese Inertial Technol 21(04):456–461
14. Tang CY, Song XC (2014) J Southeast Univ 30(04):439–444

Autonomous Navigation and Attitude Determination System Design for Micro-Nano Satellites with Limited Sensors

Chaoyu Du[1], Tao Meng[2,3](✉), Jun Wang[4], and Zhonghe Jin[2,3]

[1] School of Information Science and Electrical Engineering, Zhejiang University, Hangzhou 310027, China

[2] School of Aeronautics and Astronautics, Zhejiang University, Hangzhou 310027, China
mengtao@zju.edu.cn

[3] China Micro-Nano Satellite Research Laboratory of Zhejiang Province, Hangzhou 310027, China

[4] Operations Center, China Satellite Communications, Beijing 100190, China

Abstract. Autonomous system design has received extensive attention for orbit and attitude determination, since the traditional ground station-based orbit determination is difficult to meet the multi-satellite needs. With a combination of limited sensors, this article presents an autonomous navigation-attitude determination system for low-earth-orbit micro-nano satellites. By adding the area-to-mass ratio to state vector, the atmospheric resistance is effectively considered. By introducing an infrared earth sensor, the limitations of the orbit type and eclipse period are compensated, with the consideration of computational burden. Simulation based on ZDPS-2 satellites show the estimation accuracy of this paper is improved by 23% compared with the magnetometer/sun sensor combination, reaching 1.08 km, 1.16 m/s (RMS); whereas for an equatorial orbit, estimation accuracy remains 1.11 km (RMS). While completing navigation estimation, the system attitude pointing accuracy reaches 0.347° (RMS), which meets the basic mission requirements of micro-nano satellites.

Keywords: Navigation · Magnetometer · Earth sensor · Micro-nano satellite

1 Introduction

A growing interest has been expressed in micro-nano satellites which offer advantages such as low cost, short development period and standard with large quantities. For the attitude control system, orbital parameters are an indispensable input for any attitude determination algorithm. The traditional ground station-based orbit determination is difficult to meet the multi demands as formation flying plays a more important role in space missions. Therefore, autonomous operating system design has received extensive attention for orbit and attitude determination.

H. Ding (ed.), *Aerospace Mechatronics and Control Technology*, Springer Aerospace Technology,
https://doi.org/10.1007/978-981-16-6640-7_14

Navigation based on GPS/GNSS or star sensor, has been popular among recent research [1–3]. Whereas for micro-nano satellites which are equipped with limited sensors, it can be useful explorations on autonomous determination, if no additional sensitive components are added.

Since the magnetometer-based satellite navigation was first proposed by research team at Cornell, studies have been carried out focusing on geomagnetic navigation. The scalar magnitude of the magnetic field was introduced as measurement vector to establish the orbit determination [4]. A state estimator for a low-earth near-polar orbit reached a position accuracy of 2.5–3 km [5]. Further discussions were made by analyzing the basis of the accuracy effects, such as dynamic model, orbit types, measurement calibration, and linearization of measurement model [6].

In order to further improve system accuracy, Psiaki was among the pioneers who added sun sensor data to the geomagnetic navigation system [7], and with the measured magnetic field data of DE-2, MAGSAT, and LACE satellites [8], a batch filter and EKF algorithm reached an position error of 2.19 km, but the integrated system showed low accuracy in eclipse. Compared with EKF, a particle filter was verified and showed similar position accuracy, but better convergence rate [9]. Simplifications were introduced and testified on Jacobian calculations and polynomial model for fast magnetic field calculation [10]. Calibrations on bias and scale factors of the magnetometer were discussed [11]. By introducing a horizon sensor with UKF, a less than 500 m accuracy of position and less than 1 m/s accuracy in velocity are found with 0.1 nT magnetometer and 0.05° horizon/earth sensor [12].

The second approach for magnetometer measurements in dual estimation, on the other hand, choose the three-axis magnetic vector measurements instead of the scalar magnitude to perform the coupled navigation and attitude determination [13–16]. While the current attitude is needed when measurements are within the coordinate body system, a series–parallel hybrid determination strategy based on both magnitude and vector was proposed, in which the error covariance matrix was utilized as the switch [15].

However, there is lack of consideration fully focused on low-earth-orbit micro-nano satellites, as well as computational burden, to get fully use of the limited equipped sensors while meeting a proper system accuracy.

In this paper, based on the commonly equiped attitude devices: magnetometer, sun sensor and infrared earth sensor, an autonomous navigation-attitude determination system (low power consumption, full orbit, full time) is proposed. The solution excludes the considerations of low-earth perturbation, influence of orbit inclination, sub-system independence, as well as sensor redundancy. Finally, through simulation based on ZDPS-2 satellites, the proposed algorithm and system design is verified.

2 Design Guidelines

In this section, analyses on orbital perturbation, orbital inclination, measurement form of the magnetic field are given for low-earth-orbit satellites, which provides a design guideline for the navigation estimator and the combined determination system.

(1) **Atmospheric perturbation** cannot be ignored.
The orbital acceleration due to perturbation such as geopotential and atmospheric drag, the solar radiation pressure, and the sun and moon's gravity are calculated. The following table shows a calculation based on a 25 kg satellite with a 0.4 m^2 windward area.

As shown in Table 1, the non-spherical perturbation of the Earth has the greatest impact. For low-earth orbit satellites (400–600 km), the impact of atmospheric drag perturbation cannot be ignored. Therefore, the ballistic coefficient of the satellite is required to be estimated [6], which equals to the product of the drag coefficient and the area-to-mass ratio.

Table 1. Orbital perturbation

Perturbation	Orbit (400 km)	Orbit (1000 km)
J2	10^{-3}	10^{-3}
Other non-spherical	10^{-6}	10^{-6}
Atmospheric drag	2×10^{-6}	10^{-9}
Solar radiation pressure	10^{-9}	10^{-9}
Solar gravitational attractions	3×10^{-8}	5×10^{-8}

(2) **Orbital description** method.
The estimated state of the autonomous navigation filter can be the satellite Cartesian coordinate position, velocity vector [4, 6], the six Keplerian orbital elements [12], and other parameters forms. In contrast, the Keplerian form is easier for orbit description, and more suitable for calculation of the earth's magnetic field, but it requires a large amount of computing resources. On the other hand, the Cartesian form is more convenient for numerical integration.

(3) **Different orbital inclinations** need to be considered.
A magnetometer-based orbit determination is based on the principle that the distribution of the geomagnetic field along the orbit has adequate resolution to identify the specific orbit. Therefore, the data range of the geomagnetic field along the orbit should be large enough to allow a precise orbit reconstruction [6]. The range of the measured data is highly dependent on the type of orbit.

A verification simulation is set on a 500 km orbit (eccentricity = 0), with a 10 nT magnetometer. The filtering estimation result is shown in Fig. 1. When the orbital inclination is 90°, the position estimation accuracy is better than 5 km; while the error increases as the orbital inclination becomes closer to zero, due to the low variation of the equatorial geomagnetic field. As a result, the evaluation of a magnetometer-based navigation system, should be verified at different orbit inclinations.

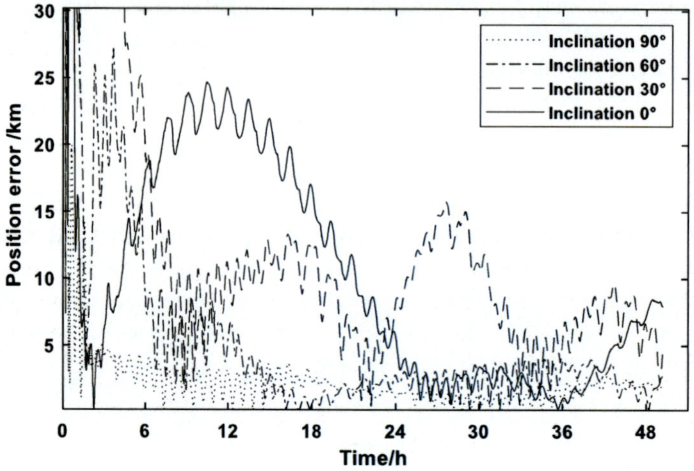

Fig. 1. Evolution of the position errors for different orbital inclinations

(4) **Convergence** of the coupled determination system.
When the three-axis magnetic vector measurements (excludes the knowledge of the attitude in the orbit determination) take place of the scalar magnitude, the coupled navigation-attitude system faces higher convergence rates. The larger estimation error of either sub system becomes, the longer it will take to be convergent. Results might be divergence as well [15].

3 System-Level Design

Based on the discussed design guidelines:

(1) In this paper, only the scalar magnitude of the geomagnetic is selected, to asure an attitude-independent orbit determination. System convergence, especially attitude determination, plays a very important role at early stage of the mission.
(2) An infrared earth imaging sensor was introduced as measurement to further enhance the observability of the commonly used magnetic/sun sensor combination; when the sun sensor becomes unavailable in eclipse, there are still continuous observations for information fusion.

The infrared earth sensor introduced in this article is shown in Fig. 2. The earth is projected onto the image plane and through image processing, the roll and pitch angle can be calculated by the following formula, with which the three-dimensional earth vector is obtained.

$$\alpha = \begin{cases} \arctan\left(\frac{Y_1 - Y_0}{X_1 - X_0}\right) & (Y_1 - Y_0) \le 0 \\ 360 - \arctan\left(\frac{Y_1 - Y_0}{X_1 - X_0}\right) & (Y_1 - Y_0) > 0 \end{cases} \tag{1}$$

$$\beta = \arctan\left(\frac{\sqrt{(X_1 - X_0)^2 + (Y_1 - Y_0)^2}}{f}\right)$$

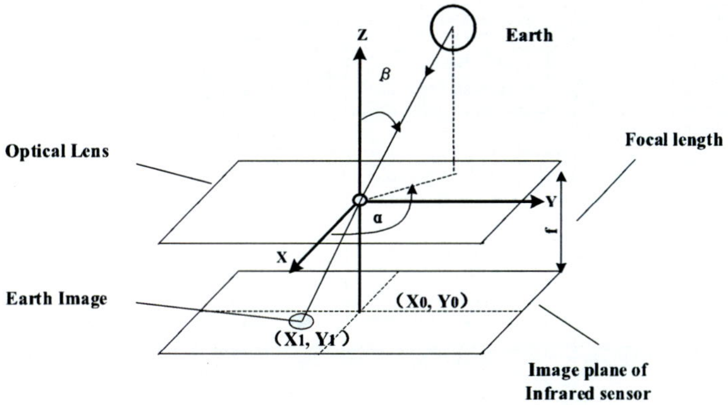

Fig. 2. Schematic principle of the static infrared earth sensor

(3) System design is shown in Fig. 3, where the fusion selector is to verify working status of the equipped sensors, they either face unavailable environment such as eclipse to sun sensor, or disabled due to self failure or under ground instructions. The status help select the latter fusion combination and algorithm.

In addition, processing of the magnetometer measurement is shown in Fig. 4, where the magnitude, and sun/earth vector angle are inputs of the navigation filter; the three-dimensional vectors are for the attitude estimator.

4 Navigation Estimator Design

The Cartesian form of coordinates is chosen for its conveniency for numerical integration. The navigation state vector is defined as follows

$$\boldsymbol{x} = \left[\boldsymbol{r}^{\mathrm{T}}\ \boldsymbol{v}^{\mathrm{T}}\ B^*\right] \tag{2}$$

where $\boldsymbol{r}$ and $\boldsymbol{v}$ are the position and velocity vectors in inertial frame, and B* is the inverse value of satellite's ballistic coefficient, which is the multiplication of the drag coefficient and the area-to-mass ratio.

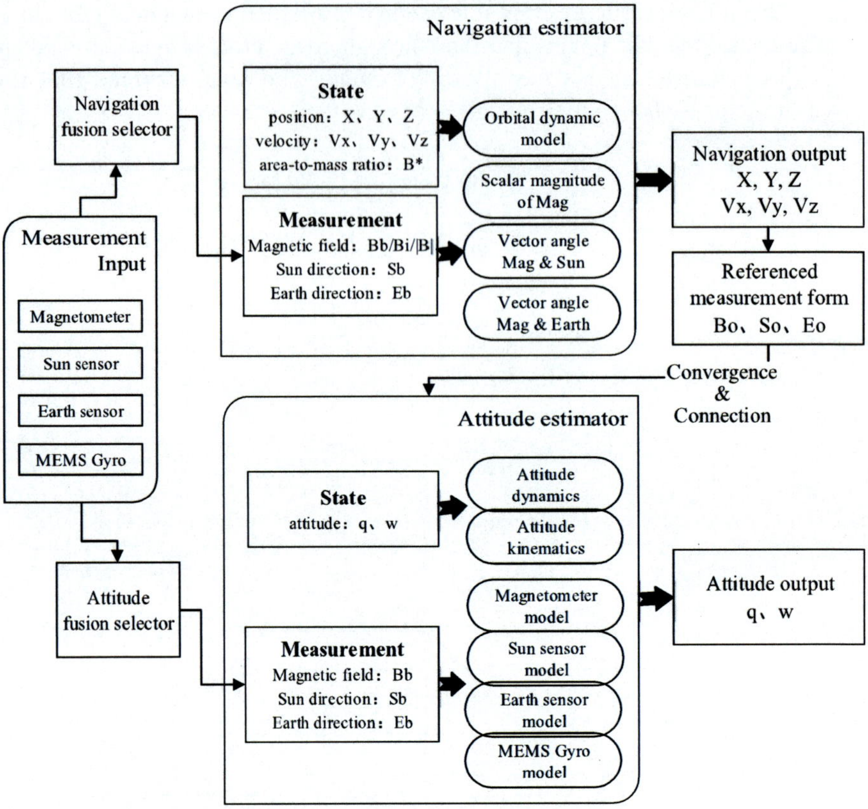

Fig. 3. System design of the autonomous navigation and attitude system

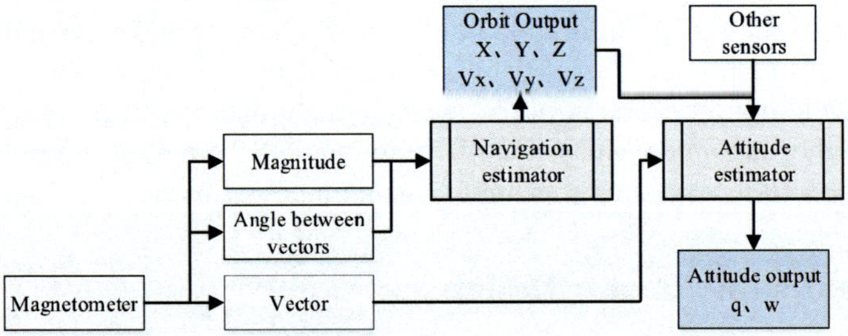

Fig. 4. Magnetometer measurement processing

4.1 Orbital Dynamic Model

The satellite, defined by multiplication of the drag coefficient and the area-to-mass ratio, and is modeled as a random walk, in which the state vector can be expressed as

$$\begin{aligned} \dot{\boldsymbol{r}} &= \boldsymbol{v} \\ \dot{\boldsymbol{v}} &= \boldsymbol{a}_{\mathrm{g}} + \boldsymbol{a}_{\mathrm{d}} + \boldsymbol{w}_1 \\ \dot{B}^* &= w_2 \end{aligned} \tag{3}$$

where $\boldsymbol{a}_{\mathrm{g}}$ denotes the geopotential acceleration, $\boldsymbol{a}_{\mathrm{d}}$ denotes the acceleration due to atmospheric drag. $\boldsymbol{w}_1$ an $\boldsymbol{w}_2$ are system process errors, which can be approximated as zero-mean Gaussian noise. For state propagation, can also be written as:

$$\dot{\boldsymbol{x}} = f(\boldsymbol{x}) + \boldsymbol{w} \tag{4}$$

where $\boldsymbol{w}$ denotes a combination of $\boldsymbol{w}_1$ and $\boldsymbol{w}_2$. Q is the system noise covariance matrix.

$$E\left(\boldsymbol{w}\boldsymbol{w}^T\right) = \boldsymbol{Q} \tag{5}$$

The Jacobian calculation of Eq. (5) can be written as

$$F(\boldsymbol{x}) = \frac{\partial f(\boldsymbol{x})}{\partial \boldsymbol{x}} = \begin{bmatrix} \boldsymbol{0}_{3\times 3} & \boldsymbol{I}_{3\times 3} & \boldsymbol{0}_{3\times 1} \\ \boldsymbol{G}_{\mathrm{r}} + \boldsymbol{D}_{\mathrm{r}} & \boldsymbol{D}_{\mathrm{v}} & \boldsymbol{D}_{\mathrm{B}} \\ \boldsymbol{0}_{1\times 3} & \boldsymbol{0}_{1\times 3} & \boldsymbol{0}_{1\times 1} \end{bmatrix} \tag{6}$$

where G_r denotes the derivative of $\boldsymbol{a}_{\mathrm{g}}$, D_r, D_v, D_B denote derivatives of $\boldsymbol{a}_{\mathrm{d}}$. Drags and J4 perturbations have been included for the orbital simulation, as well as the J2 perturbation, to obtain higher estimation accuracy with limited computational burden.

4.2 Measurement Function

Measurement model of the scalar magnitude of the geomagnetic can be written as

$$y_1 = \sqrt{\boldsymbol{B}_{\mathrm{mes}}^{\mathrm{T}}\boldsymbol{B}_{\mathrm{mes}}} \approx \sqrt{\boldsymbol{B}_{\mathrm{act}}^{\mathrm{T}}\boldsymbol{B}_{\mathrm{act}}} + n_{\mathrm{y1}} \tag{7}$$

the statistics of n_{y1} are as follows:

$$E\left(n_{\mathrm{y1}}\right) = 0, E\left(n_{\mathrm{y1}}n_{\mathrm{y1}}^{\mathrm{T}}\right) = \sigma_{\mathrm{m}}^2 \tag{8}$$

the measured dot product of the magnetic field vector and the sun direction vector y_2, and with earth direction vector y_3, can be shown as

$$\begin{aligned} y_2 &= \boldsymbol{B}_{\text{mes}}^{\text{T}} \boldsymbol{S}_{\text{mes}} \approx \boldsymbol{B}_{\text{act}}^{\text{T}} \boldsymbol{S}_{\text{act}} + n_{\text{y2}} \\ y_3 &= \boldsymbol{B}_{\text{mes}}^{\text{T}} \boldsymbol{E}_{\text{mes}} \approx \boldsymbol{B}_{\text{act}}^{\text{T}} \boldsymbol{E}_{\text{act}} + n_{\text{y3}} \end{aligned} \tag{9}$$

the statistics of n_{y2} and n_{y2} are as follows:

$$\begin{aligned} &E(n_{\text{y2}}) = E(n_{\text{y3}}) = 0, \\ &E\left(n_{\text{y2}} n_{\text{y2}}^{\text{T}}\right) = \sigma_{\text{B}}^2 + \boldsymbol{B}_{\text{mes}}^{\text{T}} \left(\boldsymbol{I} - \boldsymbol{S}_{\text{mes}} \boldsymbol{S}_{\text{mes}}^{\text{T}}\right) \boldsymbol{B}_{\text{mes}} \sigma_{\text{S}}^2 \\ &E\left(n_{\text{y3}} n_{\text{y3}}^{\text{T}}\right) = \sigma_{\text{B}}^2 + \boldsymbol{B}_{\text{mes}}^{\text{T}} \left(\boldsymbol{I} - \boldsymbol{E}_{\text{mes}} \boldsymbol{E}_{\text{mes}}^{\text{T}}\right) \boldsymbol{E}_{\text{mes}} \sigma_{\text{E}}^2 \end{aligned} \tag{10}$$

The measurement function combined can be written as

$$\begin{aligned} \boldsymbol{y} &= h(\boldsymbol{x}) + \boldsymbol{n}_{\text{y}} \\ E\left(\boldsymbol{n}_{\text{y}} \boldsymbol{n}_{\text{y}}^{\text{T}}\right) &= \boldsymbol{R} \end{aligned} \tag{11}$$

where $\boldsymbol{n}_{\text{y}}$ is a combination of n_{y1}, n_{y2} and n_{y3}. R is the measurement error covariance matrix. The Jacobian calculation of Eq. (11) can be written as

$$H(\boldsymbol{x}) = \frac{\partial h(\boldsymbol{x})}{\partial \boldsymbol{x}} = \begin{bmatrix} \frac{\partial y_1}{\partial \boldsymbol{B}_{\text{act}}} \frac{\partial \boldsymbol{B}_{\text{act}}}{\partial \boldsymbol{r}_{\text{e}}} \frac{\partial \boldsymbol{r}_{\text{e}}}{\partial \boldsymbol{r}} & \boldsymbol{0}_{1\times 4} \\ \left(\frac{\partial \boldsymbol{B}_{\text{act}}}{\partial \boldsymbol{r}_{\text{e}}} \frac{\partial \boldsymbol{r}_{\text{e}}}{\partial \boldsymbol{r}} \boldsymbol{S}_{\text{act}}\right)^{\text{T}} & \boldsymbol{0}_{1\times 4} \\ \left(\frac{\partial \boldsymbol{B}_{\text{act}}}{\partial \boldsymbol{r}_{\text{e}}} \frac{\partial \boldsymbol{r}_{\text{e}}}{\partial \boldsymbol{r}} \boldsymbol{E}_{\text{act}}\right)^{\text{T}} & \boldsymbol{0}_{1\times 4} \end{bmatrix}_{3\times 7} \tag{12}$$

where $\boldsymbol{r}_{\text{e}}$ is the position vector in geocentric coordinates.

4.3 EKF Procedure

The dynamics of orbit can be represented as follows

$$\begin{aligned} \Delta\dot{\boldsymbol{x}}(t) &= \boldsymbol{F}(t)\Delta\boldsymbol{x}(t) + \boldsymbol{G}(t)\boldsymbol{W} \\ \Delta\boldsymbol{y}(t) &= \boldsymbol{H}(t)\Delta\boldsymbol{x}(t) + \boldsymbol{V} \end{aligned} \tag{13}$$

(1) Equation of the extrapolation value can be shown as

$$\begin{aligned} \Delta\hat{\boldsymbol{x}}_{k+1|k} &= \boldsymbol{\Phi}_k \Delta\hat{\boldsymbol{x}}_{k|k} \\ \boldsymbol{\Phi}_k &= \mathbf{I}_{9\times 9} + \boldsymbol{F}(t)T \end{aligned} \tag{14}$$

where the state transition matrix $\boldsymbol{F}$ propagates the state vector in each time step T. $\boldsymbol{H}$ is known as the observation matrix, and $\boldsymbol{G}$ maps the process noise into the state vector. $\boldsymbol{x}_k$ and $\boldsymbol{x}_{k\text{-}1}$ denote the current and previous state vectors.

(2) The covariance matrix of the extrapolation error is formulated as

$$\boldsymbol{P}_{k+1|k} = \boldsymbol{\Phi}_k \boldsymbol{P}_{k|k} \boldsymbol{\Phi}_k^{\mathrm{T}} + \boldsymbol{G}_k \boldsymbol{Q}_k \boldsymbol{G}_k^{\mathrm{T}} \tag{15}$$

where $\boldsymbol{Q}_k = \begin{bmatrix} \sigma_{\mathrm{r}}^2 \mathbf{I}_{3\times3} & \mathbf{0}_{3\times3} & \mathbf{0}_{3\times1} \\ \mathbf{0}_{3\times3} & \sigma_{\mathrm{v}}^2 \mathbf{I}_{3\times3} & \mathbf{0}_{3\times1} \\ \mathbf{0} & \mathbf{0} & \sigma_{\mathrm{B}*}^2 \mathbf{I}_{1\times1} \end{bmatrix}_{7\times7}$ is the covariance matrix of system noise $\boldsymbol{W}$.

(3) Filter-gain of EKF can be shown as

$$\boldsymbol{K}_{k+1} = \boldsymbol{P}_{k+1|k} \boldsymbol{H}_{k+1}^T \left(\boldsymbol{H}_{k+1} \boldsymbol{P}_{k+1|k} \boldsymbol{H}_{k+1}^T + \boldsymbol{R}_{k+1} \right)^{-1} \tag{16}$$

where $\boldsymbol{R}_{k+1} = \begin{bmatrix} \sigma_b^2 \mathbf{I}_{3\times3} & \mathbf{0}_{3\times3} \\ \mathbf{0}_{3\times3} & \sigma_{\mathrm{s}}^2 \mathbf{I}_{3\times3} \end{bmatrix}_{6\times6}$ is the covariance matrix of measurement noise $\boldsymbol{V}$, which has diagonal elements built of the variances of magnetometer and earth sensor measurement noises σ_b, σ_s

$$\Delta\hat{\boldsymbol{x}}_{k+1|k+1} = \boldsymbol{\Phi}_k \Delta\hat{\boldsymbol{x}}_k + \boldsymbol{K}_{k+1}[\Delta \boldsymbol{y}_{k+1} - \boldsymbol{H}_{k+1} \boldsymbol{\Phi}_k \Delta\hat{\boldsymbol{x}}_{k|k}] \tag{17}$$

$$\boldsymbol{P}_{k+1|k+1} = (\mathbf{I} - \boldsymbol{K}_{k+1} \boldsymbol{H}_{k+1}) \boldsymbol{P}_{k+1|k} \tag{18}$$

With the update of the estimation and the covariance matrix of the filtering error shown in (17) and (18), the equations given above represent the Extended Kalman Filter (EKF), which fulfils the recursive estimation.

5 Simulation

Performance of the algorithm and system design was verified based on the ZDPS-2 satellites, which were launched in September 2015. Simulations are based on the equipped and limited ADCS sensors, shown in Table 2.

Table 2. ZDPS-2 ADCS sensors

Sensor	Parameter	Value	Unit
Magnetometer	Accuracy	50	nT
Sun sensor	Accuracy	0.5	deg
Earth sensor	Range	Hemispheric	
	Accuracy	0.1	deg

The following combination types, listed in Table 3, are the simulations to be taken into consideration.

Table 3. Combination and navigation-attitude mode

	Sensor	Combination	Characteristic
1	Mag only	Minimum filter	Low accuracy, with limitations
2	Mag/Sun	ZDPS-2 in orbit	Back to scenario 1 during eclipse
3	Mag/Sun/Earth	Full-sensive	The proposed system design

5.1 Navigation Simulation

Based on the original orbit of ZDPS-2 satellites, the navigation accuracy of each sensitive combination is analyzed and compared. The orbit altitude is 524 km (with eccentricity of 0.00125), and the inclination is 97.389°. As shown in Figs. 5, 6, 7 and Table 4, when the satellite area-to-mass ratio is integrated as B* and brought into the filter state equation, the estimation accuracy is improved by 12.6%, thus the perturbation caused by atmospheric drag is effectively considered for the low-orbit satellite. The full-sensitive combination (mag/sun/earth) proposed in this paper has improved estimation accuracy by 23.0% compared with the mag/sun combination, reaching 1.08 km, 1.16 m/s (RMS); due to the addition of new independent measurement, the improved observability shortens the filter convergence rate from 3.1 h to 1.4 h. Compared to the mag/earth combination, the full-sensitive combination has limited accuracy improvement, but has stronger fault tolerance and anti-interference ability.

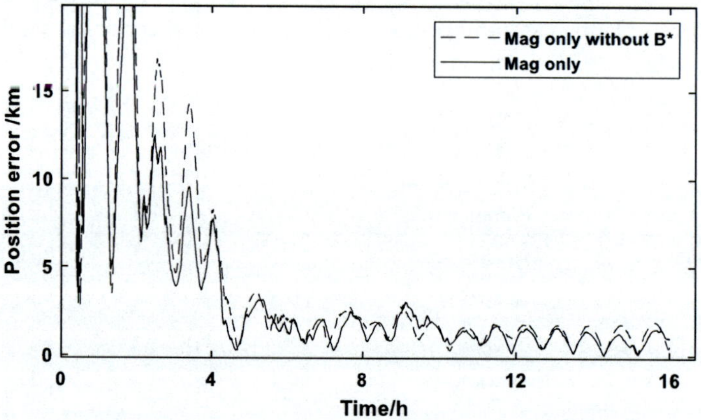

Fig. 5. Comparison with state vector B* at 97° inclination

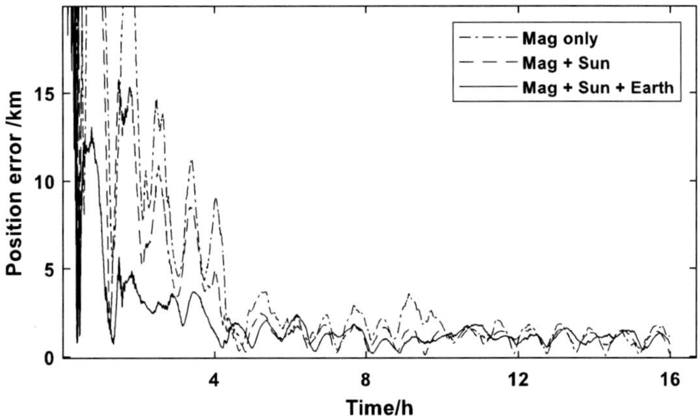

Fig. 6. Position estimation accuracy at 97° inclination

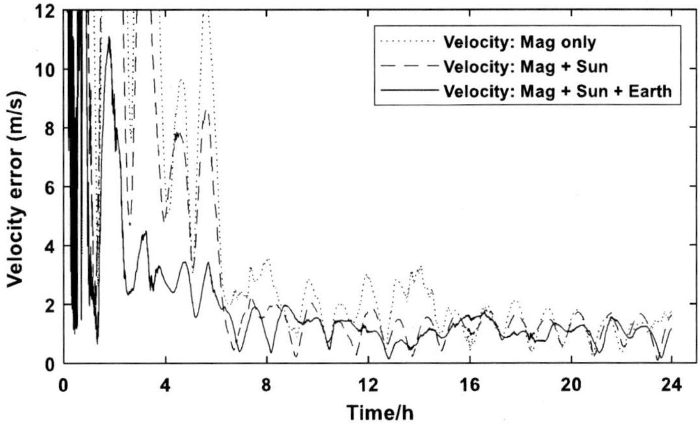

Fig. 7. Velocity estimation accuracy at 97° inclination

Table 4. Position and Velocity estimation accuracy of the navigation system

No.	Sensor	Position estimation accuracy			Velocity estimation accuracy	
		RMS/km	Max/km	Convergence (5 km)	RMS (m/s)	Max (m/s)
1	Mag (without B*)	1.637	4.378	4.2 h	1.801	3.621
2	Mag	1.454	4.161	4.2 h	1.648	3.555
3	Mag + Sun	1.328	2.868	3.1 h	1.289	2.197
4	Mag + Earth	1.105	2.376	1.5 h	1.194	2.043
5	Mag + Sun + Earth	1.078	2.319	1.4 h	1.158	1.960

The comparison simulation group carried out at the orbital inclination of 0°, as shown in Fig. 8 and Table 5. The position accuracy of the mag/sun combination becomes 10.45 km (RMS), due to the 1/3 eclipse period; however the fully-sensitive combination proposed in this paper remains 1.11 km (RMS).

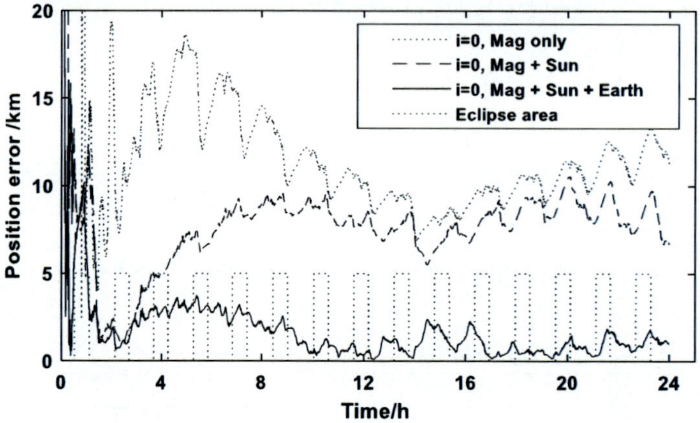

Fig. 8. Position estimation accuracy at 0° inclination

Table 5. Navigation accuracy at 0° orbit inclination

Sensor	Navigation accuracy (Position error)	
	RMS/km	Max/km
Mag	10.447	16.166
Mag + Sun	8.238	10.572
Mag + Sun + Earth	1.105	2.553

5.2 System Simulation

After the convergence of the navigation sub-system, the full-sensitive attitude filter are calculated with the prior information of the orbit estimation. System estimation error are summarized in Fig. 9 and Table 6. The accuracy of the attitude determination is improved within the full orbit range to 0.347° (RMS). The results show that the navigation and attitude determination system proposed in this paper are full orbit autonomous, with the navigation accuracy of 1.08 km and 1.16 m/s, while the three-axis pointing accuracy meets the basic mission requirements of micro-nano satellites.

6 Conclusion

A full-sensive autonomous navigation and attitude determination system is proposed. With the low-earth-orbit navigation considerations, and without the ground staion support or any other high-precision sensitive components added, the proposed algorithm and

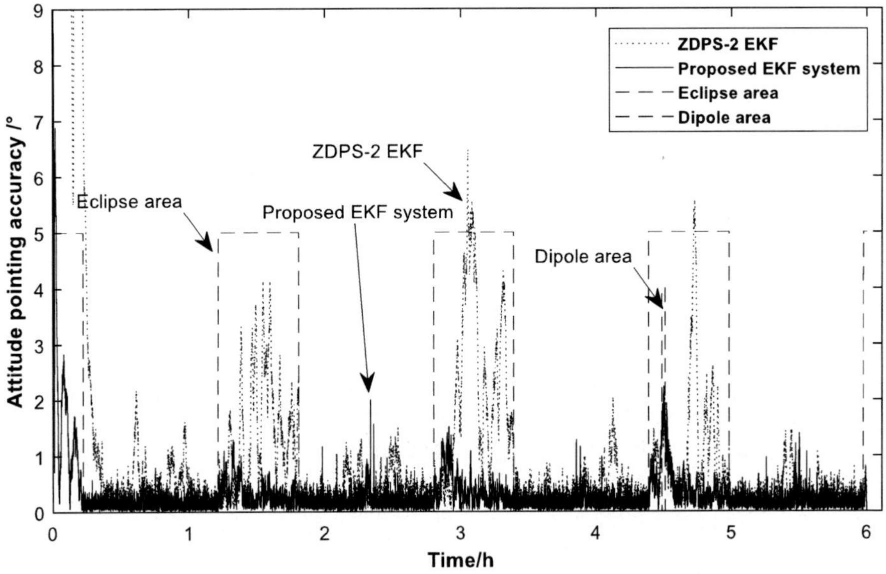

Fig. 9. Autonomous system attitude pointing accuracy

Table 6. Summary of improved filter estimation accuracy for the combined navigation and attitude system

No.	Sensor	Navigation accuracy/km (RMS)		Attitude accuracy/°(RMS)		
		RMS/km	Max/km	Sun	Eclipse	Dipole
1	Mag	1.454	4.161	2.8243		
2	Mag + Sun	1.328	2.868	1.633	/	1.633
3	Mag + Earth	1.105	2.376	0.553		/
4	Full-sensive	1.078	2.319	0.347		

system meets the basic platform requirements, and can be a useful exploration for the integration, improvement and practical design of the Micro-nano satellites navigation attitude system.

Acknowledgements. This work was supported by Primary Research and Development Plan of Zhejiang Province, key technologies of Nano remote sensing satellites, 209C05004; Zhejiang Provincial Key Laboratory of Micro-nano Satellite.

References

1. Groves PD (2015) Principles of GNSS, inertial, and multisensor integrated navigation systems. IEEE Aerosp Electron Syst Mag 30(02):26–27
2. Ning X, Wang L, Bai X, Fang J (2013) Autonomous satellite navigation using starlight refraction angle measurements. Adv Space Res 51(09):1761–1772
3. Zhang Y, Fang JC (2003) Study of the satellite autonomous celestial navigation based on the unscented kalman filter. J Astronaut 24(06):646–650
4. Psiaki ML, Huang L, Fox SM (1993) Ground tests of magnetometer-based autonomous navigation (MAGNAV) for low-earth-orbiting spacecraft. J Guid Control Dyn 16(01):206–214
5. Kim Y, Vukovich G (2014) Satellite orbit and attitude estimation using three-axis magnetometer. Int J Space Sci Eng 2(03):276–304
6. Roh KM, Park SY, Choi KH (2007) Orbit determination using the geomagnetic field measurement via the unscented Kalman filter. J Spacecr Rocket 44(01):246–253
7. Psiaki ML (1999) Autonomous low-earth-orbit determination from magnetometer and sun sensor data. J Guid Control Dyn 22(02):296–304
8. Jung H, Psiaki ML (2002) Tests of magnetometer/sun-sensor orbit determination using flight data. J Guid Control Dyn 25(03):582–590
9. Xie X, Zhang R, Zhang J (2009) Satellite autonomous orbit determination based on magnetometers and sun sensors. J Astronaut 30(03):919–923
10. Ke H, Hao W, Binjie T, Zhonghe J (2011) Pico-satellite autonomous navigation with magnetometer and sun sensor data. Chin J Aeronaut 24(01):46–54
11. Juang JC, Tsai YF, Tsai CT (2012) Design and verification of a magnetometer-based orbit determination and sensor calibration algorithm. Aerosp Sci Technol 21(01):47–54
12. Farahanifar M, Assadian N (2015) Integrated magnetometer–horizon sensor low-earth orbit determination using UKF. Acta Astronaut 106:13–23
13. Abdelrahman M, Park SY (2011) Simultaneous spacecraft attitude and orbit estimation using magnetic field vector measurements. Aerosp Sci Technol 15(08):653–669
14. Peng W, Yingchun Z (2014) J Chinese Inertial Technol 22(06):741–747
15. Xing Y, Zhang S, Cao X (2009) Determination of integrated orbit and attitude of satellite based on geomagnetic field measurements. J Harbin Inst Technol 41(07):11–15
16. Zhang T, Zheng J, Gao D (2017) A method of autonomous navigation using the magnetometer and star sensor. J Astronaut 38(02):152–158

Author Index

H. Ding (ed.), *Aerospace Mechatronics and Control Technology*, Springer Aerospace Technology,
https://doi.org/10.1007/978-981-16-6640-7

W

X

Y

Z